Long-Term Health Consequences of Exposure to Burn Pits in Iraq and Afghanistan

Committee on the Long-Term Health Consequences of
Exposure to Burn Pits in Iraq and Afghanistan

Board on the Health of Select Populations

INSTITUTE OF MEDICINE
OF THE NATIONAL ACADEMIES

THE NATIONAL ACADEMIES PRESS
Washington, D.C.
www.nap.edu

THE NATIONAL ACADEMIES PRESS   500 Fifth Street, N.W.   Washington, DC 20001

NOTICE: The project that is the subject of this report was approved by the Governing Board of the National Research Council, whose members are drawn from the councils of the National Academy of Sciences, the National Academy of Engineering, and the Institute of Medicine. The members of the committee responsible for the report were chosen for their special competences and with regard for appropriate balance.

This study was supported by Contract V101 (93) P-2136 (Task Order 19) between the National Academy of Sciences and Department of Veterans Affairs. Any opinions, findings, conclusions, or recommendations expressed in this publication are those of the author(s) and do not necessarily reflect the view of the organizations or agencies that provided support for this project.

International Standard Book Number-13: 978-0-309-21755-2
International Standard Book Number-10: 0-309-21755-5

Additional copies of this report are available from the National Academies Press, 500 Fifth Street, N.W., Lockbox 285, Washington, DC 20055; (800) 624-6242 or (202) 334-3313 (in the Washington metropolitan area); Internet, http://www.nap.edu.

For more information about the Institute of Medicine, visit the IOM home page at: **www.iom.edu**.

Copyright 2011 by the National Academy of Sciences. All rights reserved.

Printed in the United States of America

The serpent has been a symbol of long life, healing, and knowledge among almost all cultures and religions since the beginning of recorded history. The serpent adopted as a logotype by the Institute of Medicine is a relief carving from ancient Greece, now held by the Staatliche Museen in Berlin.

Cover photo by Michael Gisick. Used with permission from *Stars and Stripes*. Copyright 2010, 2011 Stars and Stripes.

Suggested citation: IOM (Institute of Medicine). 2011. *Long-term health consequences of exposure to burn pits in Iraq and Afghanistan.* Washington, DC: The National Academies Press.

*"Knowing is not enough; we must apply.
Willing is not enough; we must do."*
—Goethe

INSTITUTE OF MEDICINE
OF THE NATIONAL ACADEMIES

Advising the Nation. Improving Health.

THE NATIONAL ACADEMIES
Advisers to the Nation on Science, Engineering, and Medicine

The **National Academy of Sciences** is a private, nonprofit, self-perpetuating society of distinguished scholars engaged in scientific and engineering research, dedicated to the furtherance of science and technology and to their use for the general welfare. Upon the authority of the charter granted to it by the Congress in 1863, the Academy has a mandate that requires it to advise the federal government on scientific and technical matters. Dr. Ralph J. Cicerone is president of the National Academy of Sciences.

The **National Academy of Engineering** was established in 1964, under the charter of the National Academy of Sciences, as a parallel organization of outstanding engineers. It is autonomous in its administration and in the selection of its members, sharing with the National Academy of Sciences the responsibility for advising the federal government. The National Academy of Engineering also sponsors engineering programs aimed at meeting national needs, encourages education and research, and recognizes the superior achievements of engineers. Dr. Charles M. Vest is president of the National Academy of Engineering.

The **Institute of Medicine** was established in 1970 by the National Academy of Sciences to secure the services of eminent members of appropriate professions in the examination of policy matters pertaining to the health of the public. The Institute acts under the responsibility given to the National Academy of Sciences by its congressional charter to be an adviser to the federal government and, upon its own initiative, to identify issues of medical care, research, and education. Dr. Harvey V. Fineberg is president of the Institute of Medicine.

The **National Research Council** was organized by the National Academy of Sciences in 1916 to associate the broad community of science and technology with the Academy's purposes of furthering knowledge and advising the federal government. Functioning in accordance with general policies determined by the Academy, the Council has become the principal operating agency of both the National Academy of Sciences and the National Academy of Engineering in providing services to the government, the public, and the scientific and engineering communities. The Council is administered jointly by both Academies and the Institute of Medicine. Dr. Ralph J. Cicerone and Dr. Charles M. Vest are chair and vice chair, respectively, of the National Research Council.

www.nationalacademies.org

COMMITTEE ON THE LONG-TERM HEALTH CONSEQUENCES OF EXPOSURE TO BURN PITS IN IRAQ AND AFGHANISTAN

DAVID J. TOLLERUD (*Chair*), Professor and Chair, Department of Environmental and Occupational Health Sciences, University of Louisville, School of Public Health and Information Sciences
JOHN R. BALMES, Professor of Medicine, School of Medicine, Division of Occupational and Environmental Medicine, University of California, San Francisco
ARUNI BHATNAGAR, Director, Diabetes and Obesity Center, Professor & Distinguished University Scholar; Medicine/Cardiology, University of Louisville
EDMUND A. C. CROUCH, Senior Scientist, Cambridge Environmental, Inc.
FRANCESCA DOMINICI, Professor of Biostatistics, Department of Biostatistics, Harvard School of Public Health
ELLEN A. EISEN, Adjunct Professor of Environmental Health Sciences and Epidemiology, School of Public Health, University of California, Berkeley
MARY A. FOX, Assistant Professor, Johns Hopkins Bloomberg School of Public Health
MARK W. FRAMPTON, Professor of Medicine and Environmental Medicine, University of Rochester School of Medicine & Dentistry
PETROS KOUTRAKIS, Professor of Environmental Sciences, Department of Environmental Health, Harvard School of Public Health
JACOB McDONALD, Scientist and Director, Chemistry and Inhalation Exposure Program, Lovelace Respiratory Research Institute
GUNTER OBERDÖRSTER, Professor of Environmental Medicine, University of Rochester School of Medicine & Dentistry
DOROTHY E. PATTON, U.S. Environmental Protection Agency (Retired)
WILLIAM M. VALENTINE, Associate Professor, Department of Pathology, Vanderbilt University Medical Center
BAILUS WALKER, Professor, Environmental and Occupational Medicine, Howard University Cancer Center

Staff

ROBERTA WEDGE, Study Director
JENNIFER SAUNDERS, Program Officer (through September 2010)
DOMINIC BROSE, Associate Program Officer
CARY HAVER, Associate Program Officer
MARGOT IVERSON, Program Officer
JONATHAN SCHMELZER, Program Assistant
CHRISTIE BELL, Financial Officer
FREDRICK ERDTMANN, Director, Board on the Health of Select Populations

Reviewers

This report has been reviewed in draft form by persons chosen for their diverse perspectives and technical expertise, in accordance with procedures approved by the National Research Council's Report Review Committee. The purpose of this independent review is to provide candid and critical comments that will assist the institution in making its published report as sound as possible and to ensure that the report meets institutional standards for objectivity, evidence, and responsiveness to the study charge. The review comments and draft manuscript remain confidential to protect the integrity of the deliberative process. We wish to thank the following individual's for their review of this report:

Judy Chow, Atmospheric Sciences Division, Desert Research Institute
David Christiani, Department of Environmental Health and Department of Epidemiology, Harvard School of Public Health
David Cleverly, North Falmouth, MA
Douglas W. Dockery, Department of Environmental Health, Harvard School of Public Health
Philip K. Hopke, Center for Air Resources Engineering & Science, Department of Chemical & Biomolecular Engineering, Clarkson University
Morton Lippmann, New York Univerity, Langone Medical Center
Michael McClean, Department of Environmental Health, Boston University School of Public Health
Melissa A. McDiarmid, Department of Medicine, and Department of Epidemiology & Public Health, University of Maryland School of Medicine
Armistead G. Russell, School of Civil and Environmental Engineering, Georgia Institute of Technology
Jamie Schauer, Water Science and Engineering Laboratory, University of Wisconsin-Madison
Richard Schlesinger, Department of Biology and Health Sciences, Pace University
Kenneth R. Still, President and Scientific Director, Occupational Toxicology Associates

Although the reviewers listed above have provided many constructive comments and suggestions, they were not asked to endorse the conclusions or recommendations nor did they see the final draft of the report before its release. The review of this report was overseen by **Mark R. Cullen,** Stanford University, and **Lynn R. Goldman,**

George Washington University, School of Public Health and Health Services. Appointed by the National Research Council and Institute of Medicine, they were responsible for making certain that an independent examination of this report was carried out in accordance with institutional procedures and that all review comments were carefully considered. Responsibility for the final content of this report rests entirely with the authoring committee and the institution.

Preface

During deployment to a war zone, military personnel are exposed to a variety of environmental hazards, many of which have been associated with long-term adverse health outcomes such as cancer and respiratory disease. Many veterans returning from the current conflicts in Iraq and Afghanistan have health problems that they believe are related to their exposure to the smoke from the burning of waste in open-air "burn pits" on military bases. Particular attention has been focused on exposure to smoke, dubbed "Iraqi crud," from the open burn pit at Joint Base Balad (JBB), one of the largest U.S. military bases in Iraq.

In response to these concerns, the Department of Defense (DoD) has been conducting environmental monitoring and health studies at JBB since 2004. Screening health risk assessments, publicly released in 2008 and 2009, stated that the burn pits at JBB and other U.S. military locations in Iraq, posed an "acceptable health risk" to personnel. Nevertheless, articles in the popular press have generated widespread public concern that the JBB burn pit "may have exposed tens of thousands of troops, contractors, and Iraqis to cancer-causing dioxins, poisons such as arsenic and carbon monoxide, and hazardous medical waste." These articles, in addition to concerns expressed by military personnel and veterans and their families, helped trigger congressional hearings and legislative proposals requiring further study of burn pits. Ultimately, the Department of Veterans Affairs (VA) asked this committee to determine the long-term health effects from exposure to burn pits in Iraq and Afghanistan.

Throughout the course of its deliberations, the committee received useful information from the DoD and others that helped in the conduct of this study, such as the raw data for the air monitoring campaigns at JBB in 2007 and 2009. Unfortunately, other information that would have assisted the committee in determining the composition of the smoke from the burn pit and, therefore, the potential health effects that might result from exposure to possible hazardous air pollutants, was not available. Specifics on the volume and content of the waste burned at JBB, as well as air monitoring data collected during smoke episodes, were not available.

The committee appreciates the importance of this issue for many veterans, and it owes a tremendous thanks to the many individuals and groups who generously gave their time and expertise to share with committee members their insight into particular issues, to provide reports and data sets, and to answer queries about their work on this issue. The committee is especially grateful to the many veterans who shared their personal stories about serving on bases with burn pits, and to the many people who provided other helpful information including: Congressman Tim Bishop (D-NY); R. Craig Postlewaite, DoD Office of the Assistant Secretary of Defense (Health Affairs); Joseph Abraham, Coleen Baird Weese, Adam Deck, and Jeffrey Kirkpatrick, U.S. Army Public Health Command; Tyler Smith, National University Technology and Health Sciences Center; John Kolivosky and Scott Newkirk, U.S.

Army Institute of Public Health; William Haight and Bill Mackie, Engineering Division, The Joint Staff, Pentagon; Neema Guliani and Lisa Cody, House Committee on Oversight and Government Reform; Victoria Cassano, VA; and James Ball, American Red Cross. The committee also greatly appreciates the efforts of Rima Habre, Harvard School of Public Health, for her assistance with the positive matrix factorization modeling.

The committee is also very grateful to Roberta Wedge, who served as study director for this project, and to all of the Institute of Medicine staff members who contributed to this project: Dominic Brose, Cary Haver, Margot Iverson, Jennifer Saunders, and Jonathan Schmelzer. A thank you is also extended to William McLeod who conducted database and literature searches.

David J. Tollerud, *Chair*
Committee on the Long-Term Health Consequences of Exposure to Burn Pits in Iraq and Afghanistan

Contents

SUMMARY 1

1 INTRODUCTION 11

2 CURRENT AND HISTORICAL USES OF BURN PITS IN THE MILITARY 15

3 APPROACH TO THE TASK 23

4 EVALUATION OF AIR MONITORING DATA AND DETERMINANTS OF EXPOSURE 31

5 HEALTH EFFECTS OF AIR POLLUTANTS DETECTED AT JOINT BASE BALAD 47

6 HEALTH EFFECTS ASSOCIATED WITH COMBUSTION PRODUCTS 63

7 SYNTHESIS AND CONCLUSIONS 109

8 FEASIBILITY AND DESIGN ISSUES FOR AN EPIDEMIOLOGIC STUDY OF VETERANS EXPOSED TO BURN PIT EMISSIONS 117

Appendixes

A COMMITTEE BIOGRAPHICAL SKETCHES 129
B REVIEW OF AIR MONITORING DATA FROM JOINT BASE BALAD 133
C EPIDEMIOLOGIC STUDIES CITED IN CHAPTER 6: HEALTH OUTCOMES 139

Summary

During deployment to a war zone, military personnel are exposed to a variety of environmental hazards, such as dust, intense heat and sunlight, emissions from kerosene heaters, pesticides, and depleted uranium. Exposure to many such hazards has been associated with long-term adverse health outcomes. Many military personnel returning from the current conflicts in Iraq and Afghanistan are reporting health problems that they attribute to their exposure to emissions from the burning of waste in open-air "burn pits" on military bases. Throughout the current operations in Iraq and Afghanistan, the military has routinely used burn pits to dispose of waste.

Special controversy surrounds the burn pit used to dispose of solid waste at Joint Base Balad (JBB), near Baghdad, one of the largest military bases in Iraq and a central logistics hub for U.S. forces there. The Department of Defense (DoD) has been conducting environmental monitoring and health studies at JBB since 2004 and has previously asked the National Research Council (NRC) to review its Enhanced Particulate Matter Surveillance Program. Risk assessment studies released in May 2008 and June 2009 stated that burn pits posed an "acceptable" or "safe" health risk to personnel stationed at JBB. Nevertheless, articles in the popular press have generated widespread public concern about the pits by claiming that the JBB burn pit "may have exposed tens of thousands of troops, contractors, and Iraqis to cancer-causing dioxins, poisons such as arsenic and carbon monoxide, and hazardous medical waste." The articles helped to trigger congressional hearings and proposed legislation requiring further study of the potential health effects of exposure to burn-pit emissions on bases in Iraq and Afghanistan.

COMMITTEE'S STATEMENT OF TASK

In response to the concerns expressed by military personnel and veterans, their families, and Congress, the Department of Veterans Affairs (VA) asked the Institute of Medicine (IOM) to establish a committee to address the following statement of task:

Determine the long-term health effects from exposure to burn pits in Iraq and Afghanistan. Specifically, the committee will use the Balad Burn Pit in Iraq as an example and examine existing literature that has detailed the types of substances burned in the pits and their by-products. The committee will also examine the feasibility and design issues for an epidemiologic study of veterans exposed to the Balad burn pit.

The committee will explore the background on the use of burn pits in the military. Areas of interest to the committee might include but are not limited to investigating:

- Where are burn pits located, what is typically burned, and what are the by-products of burning;
- The frequency of use of burn pits and average burn times; and
- Whether the materials being burned at Balad are unique or similar to burn pits located elsewhere in Iraq and Afghanistan.

COMMITTEE'S APPROACH TO ITS CHARGE

IOM appointed a committee of 14 experts to carry out the study. At its first meeting, the committee decided that its approach to its task would include gathering data from the peer-reviewed literature; requesting data directly from the DoD, the VA, and other experts in the field; reviewing government documents, reports, and testimony presented to Congress; and reviewing relevant NRC and IOM reports and other literature on veterans' health issues, specific chemicals of concern, waste incineration and combustion processes, and approaches to cumulative risk assessment. The committee also held two public sessions to hear from veterans, representatives of the DoD and the VA, and other interested parties.

The committee decided that the best approach for determining the long-term health consequences of exposure to burn pit emissions was to follow the risk assessment process originally developed by the NRC in 1983, updated in 2009, and used by many federal and private organizations for protecting human and environmental health. The committee modified it to address specific issues necessitated by the statement of task. The process begins with field or laboratory measurements to characterize the nature and extent of environmental contamination. That is followed by an assessment of the magnitude of a person's or population's exposure to the contaminated environmental medium (primarily air in the case of JBB) and by a determination of the inherent toxicity of the chemical. All the information is then combined to predict the probability, nature, and magnitude of the adverse health effects that may occur from exposure.

Therefore, the committee focused first on research and data collection related to exposures and health effects reported for the populations at JBB. The committee then assessed health outcomes in other human populations potentially exposed to some of the contaminants found in burn pit emissions. On the basis of the latter information, potential exposures and health effects that might occur in the populations at JBB and other burn pit locations were assessed. Finally, the committee synthesized and summarized key findings and identified data gaps. Using the synthesis, it proposed design elements for a future epidemiologic study.

DATA COLLECTION

Several types of data were useful to the committee: information on environmental releases and concentrations of combustion products at JBB, information on possible human exposure at JBB and elsewhere, and the potential for long-term health effects of that exposure. Characterizing environmental releases and concentrations depends mainly on information on pollutant sources, qualitative and quantitative information on the pollutants present in emissions from those sources, and pollutant fate and transport in the environment. It is also necessary to identify exposed human populations and their routes of exposure.

DoD provided raw air-sampling data collected in 2007 and 2009. The raw data were useful for determining which chemicals had been analyzed for at JBB and which ones were detectable in the ambient air. All those detected were considered worth evaluating. The committee asked the DoD for information on the types and volumes of waste burned at JBB and elsewhere in Iraq and Afghanistan, but the DoD was unable to provide the committee with any information specific to the waste stream at JBB; it did, however, provide generic information on waste streams for burn pits at U.S. bases in Kosovo, Bosnia, and Bulgaria.

The committee assumed that deployed personnel were exposed to burn pit emissions mainly by inhalation, although it recognized that some ingestion and dermal exposures were possible. On the basis of the air monitoring data received from the DoD, the committee determined the adverse health effects that might be associated with the individual chemicals that were detected or that otherwise were expected to pose the greatest risk to personnel stationed at JBB. The committee relied on published summaries from diverse sources for health effects information, including IOM and NRC reports; government reports, such as those from the U.S. Environmental Protection

Agency (EPA) and the Agency for Toxic Substances and Disease Registry (ATSDR); and established databases. The committee did not re-examine the underlying data or methods for those sources but relied on them as established sources of health-effects information.

For such chemical mixtures as burn pit emissions, toxicity and other health-effects data on mixtures *themselves* are generally scarce or nonexistent. Therefore, the committee sought information on health effects in other populations that were exposed to chemical mixtures that might include at least some of the constituents of burn pit emissions. The committee gave special attention to studies of other military populations, wildland and urban firefighters, municipal incinerator workers, and residents who lived near municipal waste incinerators. Some of those studies provide data on populations that had characteristics that were similar to those of the people at JBB—for example, they were young (in 2008, 62% of active-duty personnel were 18–30 years old), healthy (deployed personnel must meet health standards), predominantly male (only 14–15% of active-duty personnel are women), and exposed to similar pollutants (chemical mixtures produced by burning).

On the basis of its data collection and literature review, the committee summarized key findings on materials burned at JBB and other military burn pit locations, health-effects data on the combustion products detected at JBB, and studies of health effects in non-Balad populations potentially exposed to similar chemicals. The committee commented on its confidence in those findings and on their utility in providing the VA with information for medical followup and future studies. The committee also considered the possible effects of coexposure to local and regional air pollution from sources other than the JBB burn pit.

The committee identified gaps in the information available on possible health effects of exposure to burn pits and discussed design and feasibility issues related to an epidemiologic study of health effects to address the gaps.

USE OF MILITARY BURN PITS

Open-air waste burning has long been used by the military when other waste-disposal options have not been available. Technologic advances in recent military conflicts mean that new items are being burned—plastic bottles and electronics, for example—and the burning of such items presents new health risks.

The uncontrolled burning of waste in pits has been the primary solid-waste management solution in Afghanistan and Iraq from the beginning of the conflicts in 2001 and 2003, respectively. The use of burn pits by the U.S. military in those countries was restricted in 2009. By December 31, 2010, their use in Iraq had gradually been phased out, but it continues in Afghanistan, where 197 burn pits were operating as of January 2011.

The DoD estimates that an average of 8–10 lb of waste is generated each day by each person in theater. On the basis of the average populations of large bases in Iraq and Afghanistan (those with more than 1,000 personnel), an average of about 30–42 tons of solid waste per day might be produced on a base. JBB, with a population that sometimes surpassed 25,000—including U.S. troops, host-nation soldiers, coalition troops, civilians, and contractors—burned perhaps 100–200 tons of waste a day in 2007. In 2009, when three incinerators were operational at JBB, about 10 tons of waste was burned daily in the pit; the burn pit ceased operation as of October 1, 2009. A 2010 Army Institute of Public Health study of burn pits in Iraq and Afghanistan reported that large bases burned waste that consisted generally of 5–6% plastics, 6–7% wood, 3–4% miscellaneous noncombustibles, 1–2% metals, and 81–84% combustible materials (further details on waste composition were not available).

In response to personnel complaints of odor, poor visibility, and health effects attributed to burn pit emissions, the U.S. Army Center for Health Promotion and Preventive Medicine (CHPPM, now the U.S. Army Public Health Command) and the Air Force Institute for Operational Health conducted ambient-air sampling and screening health-risk assessments of burn pit exposures at JBB in 2007 and again in 2009. The assessments were designed to detect potentially harmful inhalation exposures of personnel at JBB to chemicals expected to be released by the burn pit. The CHPPM reports indicated that the risk of acute health effects of all chemicals detected, except coarse particulate matter (PM), was low and that long-term health risks were "acceptable" (that is, for noncancer endpoints a hazard index of less than 1.0; for cancer endpoints a risk ranging from 1 in 10,000 to 1 in 1,000,000 or lower).

AIR-MONITORING DATA FOR JBB

The committee received raw air-monitoring data on JBB from CHPPM to use in its analysis of the expected sources and nature of air pollutants. The monitoring data were used to compare the average chemical composition of air pollution at different locations on the base and with pollution profiles for other locations around the world. Of the three monitoring locations at JBB, one was considered a background site (a mortar pit) that was usually upwind of the burn pit, and the other two locations (H-6 housing/CASF and a guard tower and transportation field) were considered to be downwind of the burn pit.

Sources of regionally and locally generated air pollutants at JBB include windblown dust, local combustion sources, and volatile evaporative emissions. The local combustion sources include the burn pit or incinerators for refuse, compression ignition vehicles, aircraft engines, diesel electric generators, and local industry and households. Volatile evaporative emissions come primarily from refueling and other fuel-management activities on the base. Each of those sources emits a complex mixture of particulate and gaseous pollutants that include volatile organic compounds (VOCs), particle-phase and vapor-phase semivolatile organic compounds, metals, and PM.

Ambient air concentrations of polychlorinated dibenzo-p-dioxins and dibenzo-p-furans (PCDDs/Fs), polyaromatic hydrocarbons (PAHs), and VOCs were measured at JBB, and the committee has used the values to estimate the effect of the burn pit on air pollution at JBB. Sampling data were evaluated for composition and concentration at each of the sites with a goal of determining differences that may be attributed to the burn pit and other known sources. The conclusions of the analyses are as follows:

- Background ambient-air concentrations of PM at JBB were high, on the average higher than U.S. air-pollution standards. The high background PM concentrations were most likely derived from local sources, such as traffic and jet emissions, and regional sources, including long-range anthropogenic emissions and dust storms, although emissions from the burn pits may have contributed a small amount of PM.
- PCDDs/Fs were detected at low concentrations in nearly all samples, and the burn pit was probably the major source of these chemicals. The toxic equivalents of the concentrations were higher than those in the United States and even in polluted urban environments worldwide, but they were below those associated locally with individual sources.
- Ambient VOC and PAH concentrations were similar to those reported for polluted urban environments outside the United States, and the major sources of those pollutants were regional background, ground transportation, stationary power generation, and the JBB airport.

Although many air pollutants were measured, some probably went unmeasured because they were not targeted. Notably, the CHPPM measurement campaigns did not include ozone, carbon monoxide, nitrogen dioxide, or sulfur dioxide—which are criteria pollutants in the United States—or other chemicals associated with combustion, such as hydrogen cyanide. The burn pit is likely to have been a source of such pollutants, so the evaluation of air-monitoring data alone cannot provide a complete picture of the potential effects of burn pit emissions. The committee appreciates the air-monitoring campaigns conducted at JBB and elsewhere in the Middle East, but flaws in the sampling design and protocols prevent a thorough understanding of the nature and sources of the air pollutants detected at JBB. The committee indicates where air-monitoring campaigns might be improved for future efforts.

The committee's conclusions suggest that the greatest pollution concern at JBB may be the mixture of regional background and local sources—other than the burn pit—that contribute to high PM. This PM, which was characterized in a different study and at different locations at JBB, consists of substantial amounts of windblown dust combined with elemental carbon and metals that arise from transportation and industrial activities. On the basis of the high concentrations in the previous studies of potentially toxic constituents of ambient PM, the air-pollution literature that focused on PM and gaseous pollutant coexposures was considered relevant to the potential morbidity of military personnel at JBB and at other sites in the Middle East.

HEALTH EFFECTS OF AIR POLLUTANTS DETECTED AT JBB

One step in the committee's analysis of the air-monitoring data was to evaluate how often a particular pollutant was detected in the samples taken. The committee decided to focus in its assessment on the 47 pollutants that were detected in at least 5% of the air-monitoring samples collected at JBB in 2007 and 2009. The committee included an additional four pollutants (1,2,4-trichlorobenzene, 1,3-dichlorobenzene, 1,3-butadiene, and 1,2-dichlorobenzene) that were detected in fewer than 5% of samples because they are expected to be present in burn pit emissions on the basis of previous experiments on combustion products released from burning waste in barrels. The committee's report summarizes the long-term health effects of 51 pollutants. Specific cancer and noncancer health-effects data on the 51 pollutants were obtained from EPA's Integrated Risk Information System (IRIS), the ATSDR Toxicological Profiles, the National Institute for Occupational Safety and Health, and the National Library of Medicine's Hazardous Substance Data Bank.

Chemicals in all three major classes of chemicals detected at JBB—PCDDs/Fs, VOCs and PAHs, and PM— have been associated with long-term health effects. A wide array of health effects has been observed in humans and animals after exposure to the specific air pollutants detected at JBB, including eye and throat irritation, organ-weight changes, histopathologic changes (for example, lesions and hyperplasia), inflammation, and reduced or impaired function. The effects have been found in many organs and systems, including adrenal glands, blood, lungs, liver, kidneys, stomach, spleen, and cardiovascular, respiratory, reproductive, and central nervous system.

The health effects of PCDDs/Fs and PM are well characterized on the basis of toxicologic, clinical, and observational epidemiologic studies. 2,3,7,8-Tetrachlorodibenzo-p-dioxin (TCDD), one PCDD congener, is classified as carcinogenic by the International Agency for Research on Cancer and as a likely carcinogen by EPA. TCDD-contaminated Agent Orange (a herbicide used in Vietnam during the war) has been associated with soft-tissue sarcoma, non-Hodgkin's lymphoma, Hodgkin disease, and chronic lymphocytic leukemia. The health effects of exposure to dioxins and dioxin-like compounds include cancer, diabetes and other endocrine system effects, immunologic responses, neurologic effects, reproductive and developmental effects, birth defects, and wasting syndrome. Such health effects as cardiovascular and respiratory morbidity and mortality and lung cancer have been associated with exposure to PM.

The health-effects data on the other pollutants detected in more than 5% of the air samples were compiled from a variety of sources that reviewed animal studies and, less often, epidemiologic investigations. The exposure conditions in many of the studies bear little resemblance to those experienced by military personnel at JBB or other base locations; the animal studies were conducted in highly controlled environments, and many of the epidemiologic studies were conducted on general populations in rural or urban areas in relatively temperate climates. Health effects associated with five or more detected chemicals include:

- Neurologic effects and reduced CNS function.
- Liver toxicity and reduced liver function.
- Cancer (stomach, respiratory, and skin cancer; leukemia; and others).
- Respiratory toxicity and morbidity.
- Kidney toxicity and reduced kidney function.
- Blood effects (anemia and changes in various cell types).
- Cardiovascular toxicity and morbidity.
- Reproductive and developmental toxicity.

The data on single-pollutant exposures have little predictive value in connection with deployed personnel at JBB or other burn pit locations where exposures are to combinations of many pollutants from both burn pits and other local and regional sources. In addition, the exact combinations of pollutants, their magnitude, and the duration of exposure of JBB personnel are unknown. There was a general lack of data on which to base an exposure assessment. Although the committee assumed that some personnel stationed at JBB worked at or very near the burn pit, there was no indication of the number of people in the vicinity of the pit, their use of personal protective equipment, how often they were in the pit, whether there was housing downwind of the pit (other than the H-6 housing) and how many people lived there and for how long, the frequency of smoke events from the pit and what was being burned that resulted in the smoke, or how long most people were stationed at JBB and their activities. In addition, it would have been helpful to have a list of other air-pollution sources, both on base and off.

Evaluating the health effects associated with a particular pollutant yields hypotheses about potential health effects of pollutant mixtures. Such hypotheses can be investigated in two ways:

- Reviewing the epidemiologic literature on health outcomes associated with exposures to burn pit emissions (recent studies on military populations) or with exposure to emissions from combustion sources similar to burn pits (firefighters and others).
- Conducting new epidemiologic investigations.

HEALTH EFFECTS ASSOCIATED WITH COMBUSTION PRODUCTS

To determine the long-term health consequences of exposure to emissions from burn pits, the committee began by reviewing health studies of military personnel exposed to the pits in Iraq and Afghanistan. However, few such studies were available, so the committee decided to approach its review of the health effects by identifying populations that were considered to be the most similar to military personnel with regard to exposures to combustion products. Two occupational groups were identified as most likely to have comparable exposures: firefighters, including those with exposure to wildland and chemical fires, and incinerator workers. Firefighters are exposed to highly complex and variable chemical mixtures. The short intermittent spikes in firefighters' exposures are likely to differ from the chronic exposures to burn-pit emissions that military personnel experience, but studies of firefighters are the best available representation of exposures to mixtures of combustion products. Household and industrial waste burned in municipal incinerators is similar to the waste reportedly burned in the pits on military bases in Iraq and Afghanistan, occupational exposures to emissions from municipal incinerators were considered to be another surrogate of exposure to burn pit emissions, although scrubbers and cleaning devices retain much of the emissions. Furthermore, because military personnel at JBB and other burn pit locations not only work on the base but live there, the committee considered the literature on people who lived near municipal incinerators to be of interest as well. Finally, studies of military personnel exposed to smoke from oil-well fires in Kuwait during the 1990–1991 Gulf War were also considered. Assessments of health effects in Gulf War veterans are particularly useful because the personnel exposed to burn pit emissions share background exposures (for example, dusty environment, vehicle exhaust, and munitions) and personnel characteristics (for example, underlying health, exposure to stressors, and general demographics) with those deployed to Operation Enduring Freedom (OEF) and Operation Iraqi Freedom (OIF).

Following the methods and criteria used by other IOM committees that have prepared reports for the *Gulf War and Health* series and the *Veterans and Agent Orange* series, the committee evaluated each epidemiologic study and designated it as a key or supporting study. The committee then discussed the weight of evidence and reached a consensus on the categories to which to assign the health outcomes considered in its report. The following categories of association were used:

- *Sufficient evidence of a causal relationship*: Evidence is sufficient to conclude that a causal relationship exists between exposure to combustion products and a health outcome in humans. The evidence fulfills the criteria for sufficient evidence of a causal association and satisfies several of the criteria used to assess causality: strength of association, dose–response relationship, consistency of association, temporal relationship, specificity of association, and biologic plausibility.
- *Sufficient evidence of an association*: Evidence is sufficient to conclude that there is a positive association. That is, a positive association has been observed between exposure to combustion products and a health outcome in human studies in which bias and confounding could be ruled out with reasonable confidence.
- *Limited/suggestive evidence of an association*: Evidence is suggestive of an association between exposure to combustion products and a health outcome in humans, but it is limited because chance, bias, and confounding could not be ruled out with reasonable confidence.

- *Inadequate/insufficient evidence to determine whether an association exists*: The available studies are of insufficient quality, consistency, or statistical power to permit a conclusion regarding the presence or absence of an association between exposure to combustion products and a health outcome in humans.
- *Limited/suggestive evidence of no association*: Several adequate studies, covering the full range of levels of exposure that humans are known to encounter, are mutually consistent in not showing a positive association between exposure to combustion products and a health outcome. A conclusion of no association is inevitably limited to the conditions, levels of exposure, and length of observation covered by the available studies. In addition, the possibility of a small increase in risk at the levels of exposure studied can never be excluded.

The studies discussed by the committee have limitations and uncertainties—some common to epidemiologic studies in general, and some specific to studies of working populations. The limitations and uncertainties include the healthy worker effect, exposure misclassification, lack of information on confounders, inadequate statistical power, disease misclassification, and publication bias.

On the basis of a review of the epidemiologic literature, the committee concluded that there is inadequate/insufficient evidence of an association between exposure to combustion products and cancer, respiratory disease, circulatory disease, neurologic disease, and adverse reproductive and developmental outcomes in the populations studied. However, there is limited/suggestive evidence of an association between exposure to combustion products and reduced pulmonary function in the populations studied. The committee further concluded that additional study of health effects specifically in OEF and OIF veterans is necessary. The research considered by the committee is a best attempt to use currently available information on combustion products to extrapolate to exposures of military personnel stationed at JBB; however, because of differences in exposure, stress, population characteristics, access to medical care, and monitoring of health, the results in firefighters, incineration workers, and people living near incinerators may not be generalizable to military personnel exposed to emissions from burn pits.

SYNTHESIS

The committee based its conclusions regarding the long-term health consequences of exposure to emissions from burn pits in Iraq and Afghanistan on three sources of information: data on air monitoring at JBB, health-effects information on chemicals detected in more than 5% of the air-monitoring samples at JBB, and health-effects information on populations considered to be surrogates of military personnel exposed to combustion products from burn pits: firefighters, municipal incinerator workers, residents who live near incinerators, and veterans of the 1990–1991 Persian Gulf War who were exposed to smoke from oil-well fires.

The air-monitoring data suggest that the pollutants of greatest concern at JBB may be the mixture of chemicals from regional background and local sources—other than the burn pit—that contribute to high PM. The PM consists of substantial amounts of windblown dust combined with elemental carbon and metals that arise from transportation and industrial activities. On the basis of the high concentrations and the observation of potentially toxic constituents in ambient PM in previous studies, the air-pollution literature on PM and gaseous-copollutant exposures is considered relevant to the potential morbidity of military personnel at JBB and at other sites in the Middle East.

In light of its assessment of health effects that may result from exposure to air pollutants detected at JBB and its review of the literature on long-term health effects in surrogate populations, the committee is unable to say whether long-term health effects are likely to result from exposure to emissions from the burn pit at JBB. However, the committee's review of the literature and the data from JBB suggests that service in Iraq or Afghanistan—that is, a broader consideration of air pollution than exposure only to burn pit emissions—might be associated with long-term health effects, particularly in highly exposed populations (such as those who worked at the burn pit) or susceptible populations (for example, those who have asthma), mainly because of the high ambient concentrations of PM from both natural and anthropogenic, including military, sources. If that broader exposure to air pollution turns out to be sufficiently high, potentially related health effects of concern are respiratory and cardiovascular effects and cancer. Susceptibility to the PM health effects could be exacerbated by other exposures, such as stress, smoking, local climatic conditions, and coexposures to other chemicals that affect the same biologic or chemical processes. Again, further information on which to base an exposure assessment would have been helpful.

Specifically, none of the individual chemical constituents of the combustion products emitted at JBB appears to have been present at concentrations likely to be responsible for the adverse health outcomes studied in this report. However, the possibility of exposure to mixtures of those chemicals raises the potential for health outcomes associated with cumulative exposure to combinations of the constituents in burn pit emissions. As a preliminary step toward understanding possible long-term health effects of multiple contaminants or cumulative exposure, the committee looked at all the detected pollutants and the target organs or specific effects associated with them. Because a specific adverse health outcome may be influenced by several chemicals in a mixture, the overall effect of the mixture may be to increase the likelihood or severity of the outcome. Many of the chemicals detected at JBB are known to produce similar health effects—for example, anemia, reduced liver function, and birth defects—or to act on the same organs or organ systems, such as the liver, kidneys, central nervous system, and cardiovascular system.

NEW STUDY DESIGN AND FEASIBILITY

The available epidemiologic studies considered by the committee for this report are inconsistent in quality, were conducted with varied methodologic rigor, and had considerable variations in study design and sample size. The CHPPM report that described several health outcomes in personnel stationed at JBB is the first step in addressing some of those issues, but the period of followup was too short to detect long-term health effects. In addition, the difficulties in determining exposure are apparent, and better exposure assessment is critical if one is to attribute adverse health outcomes to burn pit exposures rather than to exposures common to war and desert environments.

Most critically, the database on the nature and extent of exposure to combustion products is incomplete. Given an awareness of the data gaps and analytic limitations in the studies reviewed for this report, the committee recommends a prospective study of the long-term health effects of exposure to burn-pit emissions in military personnel deployed at JBB. To determine the incidence of chronic diseases or cancers that have long latency, people must be followed for many years. Thus, it is critical that observation for health effects begin at first deployment to JBB and continue long enough after active duty is completed to detect latent health effects. The committee recommends that pilot studies be conducted to address issues of statistical power and to develop design features for specific health outcomes. It is important to note that once a prospective cohort infrastructure has been established, multiple health outcomes can be studied in the cohort over time. Intermediate outcomes on the pathway to the development of chronic diseases can also be studied in a serial manner.

To characterize exposures to the complex mixture of burn pit emissions in light of the presence of other sources of air pollutants in the ambient environment, the committee recommends a tiered approach. The three tiers of the recommended study are characterized by the decreasing specificity of exposure and would answer different research questions, as follows:

- **Tier 1:** *Did proximity to burn pit operations at JBB increase the risk of adverse health outcomes?*

    Assess individual exposure to JBB burn pit emissions (for example, low, medium, and high) on the basis of dates of deployment, duties on base, and location of housing relative to the burn pit, taking account of wind-dispersion models. The exposure effect can be assessed by comparing subgroups with more and less exposure among all potentially exposed people, that is, those stationed at JBB during the period of full burn-pit operation (2003–2007). The use of soil dioxin concentrations at various locations at JBB should be considered as a potential marker of exposure to burn pit emissions.

- **Tier 2:** *Did installation of incinerators at JBB reduce the incidence of disease or intermediate outcomes (for example, the rate of lung-function decline or the increase in intima–media thickness)?*

    Assess exposure (yes or no) to the JBB burn pit according to date of initial deployment. This approach considers the installation of incinerators during 2008–2010 to replace the burn pit as an intervention. Chronic health outcomes in those deployed before and those deployed after the burn pit was shut down can be compared, and the increased use of incinerators over 2 years can be factored in.

- **Tier 3:** *Did deployment at JBB during full burn pit operation increase the risk of adverse health outcomes compared with deployment elsewhere in Iraq or Afghanistan or with no deployment?*

    Assess exposure (yes/no) to the total JBB environment, recognizing that the burn pit emissions occurred in the presence of PM and other air pollutants from other sources. This broad definition of exposure can be assessed by comparing the health experience of military personnel deployed at the JBB during the period of burn pit operation to that of military personnel deployed to Iraq and Afghanistan at locations without a burn pit or that of military personnel not deployed to the Middle Eastern theatre during the same time. This approach was used to conduct the short-term health studies described in Chapter 6. Although there are limitations to this approach, it may be possible to find an appropriate unexposed comparison group—preferably another deployed population unexposed to burn pits but exposed to PM and other chemicals identified at JBB from other sources. The recommendation for a nondeployed comparison group is based on the committee's judgment that pollution in the region from sources other than burn pits may pose greater health risks than burn pit emissions.

1

Introduction

During deployment, military personnel are exposed to a variety of environmental hazards, such as dust, intense heat and sunlight, kerosene heaters, pesticides, and depleted uranium. Exposure to some such hazards has been associated with long-term adverse health outcomes. Many U.S. military personnel returning from the current conflicts in Iraq and Afghanistan are reporting health problems that they contend are related to their exposure to emissions from the burning of waste in open-air "burn pits" on military bases. Throughout the military operations in Iraq and Afghanistan, burn pits have been used to dispose of all types of waste.

Military field operations generate large quantities of waste that must be disposed of. It is estimated that about 8–10 pounds of solid waste is generated per person per day at the bases in Iraq and Afghanistan, although this number can vary depending on the base and its population (Faulkner 2011). Historically, the U.S. military has established open-air waste burning sites when hauling trash to appropriate disposal sites or the sanitary discharge of latrine wastes are not available options. A wide range of refuse can be burned, from food and human waste to packaging and equipment, as well as materials abandoned by the enemy. Current Department of Defense (DoD) regulations permit the use of burn pits only until better disposal options such as incinerators are available and operational (AFIOH undated). Nevertheless, burn pit use continues to be widespread at U.S. military bases in Afghanistan, although they are being phased out in Iraq.

Special controversy surrounds the open burn pit used to dispose of solid waste at Joint Base Balad (JBB), near Baghdad, one of the largest military bases in Iraq and a central logistics hub for U.S. forces there. On the cover of this report is a photograph of the burn pit at JBB, taken for the *Stars and Stripes* newspaper. The DoD has been conducting environmental monitoring and health studies at JBB since 2004 and a health screening study, released in May 2008, stated that burn pits posed an "acceptable" health risk (Taylor et al. 2008). Nevertheless, an *Army Times* article published in October 2008 generated widespread public concern about the pits by reporting that the JBB burn pit "may have exposed tens of thousands of troops, contractors, and Iraqis to cancer-causing dioxins, poisons such as arsenic and carbon monoxide, and hazardous medical waste" and emphasizing claims that such exposures resulted in reported illnesses among troops. This article and subsequent ones helped to trigger congressional hearings and legislative proposals requiring further study of the potential health effects that might arise from exposure to burn pits on U.S. military bases in Iraq and Afghanistan.

JBB, also known as Logistic Support Area Anaconda, is a 10 square-mile forward operating base located approximately 40 miles north of Baghdad. The military population of the base fluctuated over the years when the burn pit was in operation from an average of 241 service members within a 5-mile radius of JBB in 2003 to an average of 10,430 service members in 2009; the average population of U.S. service members peaked at over 15,000 in 2007 (Steve Halko, Defense Manpower Data Center, personal communication, August 25, 2010). An unknown number of coalition forces and civilian contactors were also on the base. Because of its large population, the scale of the Balad burn pit was also large, with estimates of the amount of waste burned ranging from about 2 tons per day early in its operation in 2003 to 200 tons of waste being burned daily in 2007 (Taylor et al. 2008; USAPC 2010). The burn pit ceased operating in late 2009.

No inventory of the items burned in the pit was made available to the committee, but the refuse was reported to include a wide variety of materials that could produce potentially hazardous emissions. Among these substances were plastics, metal cans, rubber, chemicals (paints, solvents), petroleum, munitions, and wood waste. In addition, JP-8 jet fuel was used as an accelerant for the fire (Taylor et al. 2008). Anecdotal accounts mention specifically the burning of plastic water bottles, food waste, human waste, and munitions, but ordnance does not appear to have been burned deliberately in the JBB burn pit.

Uncontrolled open-air burning does not completely burn the wastes, and military documents, eyewitness accounts, and publicly available photographs and videos confirm that at JBB and at other bases smoke plumes rose from burn pit areas, and at times smoke blew over the base and into living areas (Taylor et al. 2008). Open air pit burning at JBB generated complaints as early as 2003 (CHPPM undated).

In 2007, in response to the concerns of active-duty military personnel about potential hazardous inhalation exposures at JBB and other installations with burn pits, the U.S. Army Center for Health Promotion and Preventative Medicine (CHPPM, now called the U.S. Army Public Health Command) and the U.S. Air Force Institute for Operational Health (AFIOH) began a formal screening health risk assessment at JBB (Taylor et al. 2008). The study collected air samples from January through April of 2007 at several locations around the base, and the samples were analyzed for many of the chemicals expected to be emitted during trash burning. The objective was to assess the potential for adverse health effects to personnel stationed at the base who might be exposed to such chemicals. The CHPPM report (released in May 2008) found risks for cancer or noncancer health effects of concern that could be attributed to exposure to the air pollutants detected at JBB to be "acceptable" (that is, a cancer risk between 1 in 10,000 and 1 in 1,000,000 or lower, and a hazard index less than one). However, it should be noted that many air pollutants were not measured at JBB including U.S. National Ambient Air Quality Standard priority pollutants such as nitrogen and sulfur oxides, ozone, and carbon monoxide. Follow-up sampling in the same locations occurred in the fall of 2007 after two incinerators were installed, and subsequently in 2009 just before complete closure of the burn pit. Screening health risk assessments based on those sampling campaigns also found the risks from exposure to the air pollutants to be "acceptable" for cancer risks, with some potential for short-term, reversible non-cancer effects, and a "moderate" operational risk from particulate matter (CHPPM and AFIOH 2009; Taylor et al. 2009; USAPHC 2010).

Members of Congress also became interested in military burn pit use and safety. Bills were introduced in 2009 and 2010 to sharply curtail the use of open-air burn pits and establish a medical surveillance system to identify veteran health effects attributed to exposure to the burning of solid waste. HR 2647, the National Defense Authorization Act for Fiscal Year 2010, prohibits the use of burn pits for hazardous prohibits the use of burn pits for hazardous and medical waste except in cases where there is no alternative,[1] and the act requires the DoD to take several actions including: reporting to Congress regularly whenever burn pits are used; developing a plan for alternatives to burn pits; assessing existing medical surveillance programs of burn pits exposure and making recommendations to improve them; and studying the effects of burning plastics in open pits and evaluating the feasibility of prohibiting the burning of plastics. In 2009, congressional hearings on the proposed bill included testimony from both military officials and veterans groups and focused on the CHPPM screening study, with DoD and Department of Veterans Affairs (VA) officials emphasizing the study's conclusion that JBB exposures fell within military exposure guidelines and U.S. Environmental Protection Agency values for acceptable risk.

---

[1]Hazardous wastes in the National Defense Authorization Act uses the definition in the 2002 Solid Waste Disposal Act, Section 1004(5) to mean "a solid waste, or combination of solid wastes, which because of its quantity, concentration, or physical, chemical, or infectious characteristics may—(A) cause, or significantly contribute to an increase in mortality or an increase in serious irreversible, or incapacitating reversible, illness; or (B) pose a substantial present or potential hazard to human health or the environment when improperly treated, stored, transported, or disposed of, or otherwise managed" (http://epw.senate.gov/rcra.pdf; accessed August 23, 2011).

Veterans and individual medical and environmental professionals who served in Iraq testified about the presence of noxious smoke on the base and attributed a range of medical problems to smoke from burn pits, including asthma, joint pain, cancer, vomiting and nausea, burning lungs, and Parkinson's disease. In addition, medical and environmental personnel testified about the increased respiratory symptoms reported by personnel on bases in Iraq and by returning veterans seeking medical treatment stateside (U.S. Congress 2009a,b).

STATEMENT OF TASK

In response to the concerns expressed by service members, their families, and Congress, the VA asked the Institute of Medicine (IOM) to examine the long-term health consequences of exposure to burn pits in Iraq and Afghanistan. The IOM established a committee that was given the following statement of task:

> Determine the long-term health effects from exposure to burn pits in Iraq and Afghanistan. Specifically, the committee will use the Balad Burn Pit in Iraq as an example and examine existing literature that has detailed the types of substances burned in the pits and their by-products. The committee will also examine the feasibility and design issues for an epidemiologic study of veterans exposed to the Balad burn pit.
>
> The committee will explore the background on the use of burn pits in the military. Areas of interest to the committee might include but are not limited to investigating:
>
> - Where are burn pits located, what is typically burned, and what are the by-products of burning;
> - The frequency of use of burn pits and average burn times; and
> - Whether the materials being burned at Balad are unique or similar to burn pits located elsewhere in Iraq and Afghanistan.

COMMITTEE'S APPROACH TO ITS CHARGE

IOM appointed a committee of 14 members with expertise in occupational and environmental health, toxicology, exposure assessment and modeling, epidemiology, clinical medicine, and biostatistics to carry out the study. At its first meeting, the committee decided that its approach to gathering information would include considering data from the peer-reviewed literature; gathering data directly from the DOD and the VA and other experts in the field; reviewing government articles, reports, and testimony presented to Congress; and reviewing relevant National Research Council (NRC) and IOM reports on veterans health issues, specific chemicals of concern, waste incineration and combustion processes, and approaches to cumulative risk assessment. In addition, the committee held two public sessions to hear from veterans, representatives from the DoD and the VA, and other knowledgeable parties. Discussion with staff of the House Committee on Oversight and Government Reform in October 2010 on the congressional investigation of burn pits also helped to inform the committee's understanding of the available documentation on military burn pits in Iraq and Afghanistan.

Although the committee conducted extensive searches of the peer-reviewed literature in its attempts to understand health consequences of exposure to burn pit smoke and emissions in Iraq and Afghanistan, there was a paucity of information published in the peer-reviewed literature related specifically to health effects from such burning. In the absence of published data on in-theater burn pit emissions, the committee reviewed reports from the DoD and published literature on emissions from all types of open burning activities. The committee also requested additional data from the DoD on the environmental monitoring conducted at JBB for the screening reports. Although the committee asked the DoD for information on the types and volumes of waste burned at JBB or elsewhere in Iraq and Afghanistan, the DoD was unable to provide the committee with any specific information but it did provide generic information on waste streams for burn pits at U.S. installations in Kosovo, Bosnia, and Bulgaria (Faulkner 2011).

ORGANIZATION OF THE REPORT

This report includes the committee's assessment of the potential long-term health effects of exposure to burn pit smoke in Iraq and Afghanistan along with feasibility and design issues for an epidemiologic study of veterans exposed to the Balad burn pit. Chapter 2 provides information on the current and historical use of burn pits in the military and the history of concerns about possible health risks related to exposure to burn pit smoke. The committee's approach to the task is described in Chapter 3. An evaluation of air monitoring data from JBB and determinants of exposure are discussed in Chapter 4. The long-term health effects associated with the chemicals identified in the emissions from the burn pit at JBB are discussed in Chapter 5. Chapter 6 summarizes the potential long-term health effects associated with exposure to combustion products for other populations such as firefighters. Chapter 7 summarizes the conclusions reached in Chapters 4 through 6 and synthesizes that information into an evaluation of the long-term health effects that might be associated with exposure to burn pit emissions in Iraq and Afghanistan. Chapter 8 describes feasibility and design issues for an epidemiologic study of veterans exposed to burn pit emissions.

REFERENCES

AFIOH (U.S. Air Force Institute for Operational Health). Undated. *Open pit burning general facts and information*. Brooks City-Base, TX: U.S. Air Force Institute for Operational Health.

CHPPM (U.S. Army Center for Health Promotion and Preventative Medicine) and AFIOH. 2009. *Addendum 2. Screening health risk assessment burn pit exposures, Balad Air Base, Iraq, May 2008*. USACHPPM Report No. 47-MA-08PV-08/AFIOH Report No. IOH-RS-BR-TR-2008-0001. Aberdeen Proving Ground, MD: U.S. Army Center for Health Promotion and Preventive Medicine. August.

CHPPM. Undated. *Just the facts: Balad burn pit*. Aberdeen Proving Ground, MD: U.S. Army Center for Health Promotion and Preventive Medicine.

Faulkner, W. M. 2011. *Exposure to toxins produced by burn pits: Congressional data request and studies*. Memorandum for the assistant secretary of defense for health affairs. Washington, DC: Joint Staff. March 28, 2011. Enclosure: ASD(HA) Memorandum, 17 Feb 11. Response to ASD(HA) Request for Information.

Taylor, G., V. Rush, A. Deck, and J. A. Vietas. 2008. *Screening Health Risk Assessment Burn Pit Exposures, Balad Air Base, Iraq and Addendum Report*. IOH-RS-BR-TR-2008-0001/USACHPPM 47-MA-08PV-08. Brooks City-Base, TX: Air Force Institute for Operational Health and U.S. Army Center for Health Promotion and Preventative Medicine. May.

U.S. Congress, Senate, Committee on Veterans' Affairs. 2009a. *Airway injury in U.S. soldiers following service in Iraq and Afghanistan*. 111th Cong., 1st Sess. October 8.

U.S. Congress, Senate, Democratic Policy Committee. 2009b. *Are burn pits in Iraq and Afghanistan making our soldiers sick?* 111th Cong., 1st Sess. November 6.

USAPHC (U.S. Army Public Health Command). 2010. *Screening health risk assessments, Joint Base Balad, Iraq, 11 May–19 June 2009*. Aberdeen Proving Ground, MD: U.S. Army Public Health Command. July.

2

Current and Historical Uses of Burn Pits in the Military

Military engagements have always generated garbage. Troops on the move, or in temporary camps, create all the waste that human communities typically do—food remains, latrine waste, and all the other assorted detritus of living. Military populations produce additional types of waste as well, some of them hazardous to human health and the environment, and many of them in large volumes. Electronics, weapons and munitions, biological waste from combat and medical care, plastic devices of various kinds, rubber tires—all must be disposed of, often in situations where the usual waste management systems such as land-filling, recycling, and incineration are not viable options. In situations where there is no established system to safely and efficiently dispose of these waste products, they can quickly become health and safety hazards.

Open-air waste burning is one option that has long been used by the military when other options are not available. The types of items being burned, however, have changed over time. Technological advances have meant that in recent military conflicts there are new items being burned—plastic bottles and electronics, for example—and the burning of these items present new types of health hazards.

Burn pits are designated areas, sometimes excavated to create depressions and mounds of earth to aid in concentrating the burning, used to dispose of a wide variety of waste products. This chapter documents some of what is known about burn pits use by the military in the current conflicts in Iraq and Afghanistan.

HISTORY OF MILITARY WASTE MANAGEMENT

In past conflicts, poorly managed waste has contributed to the spread of infectious disease, a substantial cause of mortality and morbidity in military populations as well as a contributing factor in the contamination of drinking water. Thus proper sanitation has been historically and remains today the most important issue driving military waste management practices (Weese and Abraham 2009).

Methods for military waste management have often adapted current civilian waste management practices for use in military contexts. Beginning in the 19th century, the recognition that poor waste disposal could increase disease transmission led cities and towns to isolate, remove, and decontaminate waste. For solid waste, there were a number of common options including dumping (in nearby waterways or on land), land-filling, reduction (reducing the production of waste), compaction (reducing the volume of waste), recycling, and burning (Melosi 2005).

Open-air burning greatly reduces the volume of waste, but because it presents a fire danger and produces noxious fumes, by the early 20th century incinerators became widely used in American cities. Open-air burning remained in use in rural areas for much of the 20th century, but in recent years many state environmental protection offices, in collaboration with the U.S. Environmental Protection Agency (EPA), have enacted bans on backyard trash burning owing to air pollution concerns (EPA 2011). Concerns over the health consequences of air pollution have led to a preference for land-filling and recycling over incineration in urban areas, although incinerators continue to be used in some places (Melosi 2005).

In military contexts, open-air burning has proven to be of enduring utility in certain situations. Among its advantages are that it is less expensive than other forms of waste disposal to set up and manage and it can be operational in a short time at a newly constructed camp. Incinerators, in comparison, are relatively expensive in terms of both human labor (trash must be sorted to control moisture content and to remove hazardous materials) and upfront investment in equipment. As a result, incinerators are often considered impractical for use in rapidly changing combat situations. Other types of waste management, such as land-filling or recycling may not be feasible in many overseas operations.

BURN PITS IN IRAQ AND AFGHANISTAN

The burning of waste in pits has been the primary solid waste management solution in Iraq and Afghanistan from the beginning of the conflicts in 2003 and 2001, respectively. The use of burn pits by the U.S. military in these countries was restricted by U.S. law in 2009, and as of December 31, 2010, their use has been gradually phased out in Iraq but continues in Afghanistan (GAO 2010; DoD 2011). The exact number of burn pits in use in Iraq and Afghanistan is difficult to determine because of the constant fluctuations in the number of operational bases. The use of burn pits also varies depending on the size of the base; some bases might have a burn pit operating one day a week, others 24 hours a day, seven days a week, depending on the activities on the base and the size of its population.

This committee requested information from the Department of Defense (DoD) on the location of burn pits in Iraq and Afghanistan and their frequency of use and average burn times. The DoD reported that although data were not available for all sites, as of November 2009, in Iraq burn pits were operating at 14 out of the 41 existing small military sites (defined as housing less than 100 U.S. service members), 30 of the 49 medium-size sites (between 100 and 1,000 service members), and 19 of the 25 large sites (more than 1,000 service members) (DoD 2011).

The number of burn pits used in Iraq has declined in response to the 2009 regulations, and a 2010 Government Accountability Office (GAO) study of open-air pit burning in Iraq and Afghanistan listed only 22 burn pits in use in Iraq in August 2010 (GAO 2010). The use of burn pits in Afghanistan, however, continues and in January 2011, 126 out of the 137 small sites, 64 of the 87 medium-size sites, and 7 of the 18 large military installation sites in Afghanistan had operating burn pits (DoD 2011).

The committee also requested information from the DoD on the types and volumes of materials burned in pits in Iraq and Afghanistan. The DoD responded that no data were available on the volumes of trash burned, but estimated that, on average, 8–10 pounds of waste were generated each day by each person in theater. Based on the average populations of large bases of Iraq and Afghanistan, this would mean that these large bases would produce approximately 60,000–85,000 pounds of solid waste per day (DoD 2011). The DoD characterized the waste burned at Joint Base Balad (JBB; also called Balad Air Force Base and Logistic Support Area Anaconda) in Iraq as "municipal waste" (Taylor et al. 2008). The committee was provided with a summary of the results of a 2010 study by the Army Institute of Public Health on burn pits in Iraq and Afghanistan as well as several older studies of waste production at American bases in the Balkans (Faulkner 2011). The Army study reported that large bases in Iraq and Afghanistan burned waste with the general composition of 5–6% plastics, 6–7% wood, 3–4% miscellaneous noncombustibles, 1–2% metals, and 81–84% combustible materials (further details on waste composition were not available). The committee notes a considerable variation in the types of materials burned at each base, and the DoD states that the efforts to segregate waste and recycle materials "are often inconsistent and dependent on the waste being received and the personnel available to perform the waste segregation/recycling duties" (Faulkner 2011).

In addition, bases differ in terms of burn pit oversight. At some bases the burn pit is managed directly by DoD, while at many others the burn pit is managed by civilian contractors through the Logistics Civil Augmentation Program, which is the Army program for logistics support using civilian contractors. Some bases in Iraq and Afghanistan are able to use local contractors to remove waste from bases and operate off-site burn pits.

For all of these reasons, the committee finds that an assessment of one burn pit on a U.S. military base is unlikely to be applicable to all, or most, burn pits on U.S. military bases in Iraq and Afghanistan. Burn pit activities and thus their emissions will, in most cases, need to be considered on an individual basis.

The Burn Pit at Joint Base Balad

The burn pit at JBB has been in operation at least since 2003 when the base was established at a former air base of the Iraqi military, originally built in the 1980s (Taylor et al. 2008). The burn pit operation was used to burn the trash generated by the base population as well as refuse left by the Iraqi military when it abandoned the base (personal communication, William Haight, Engineering Division, Joint Staff, DoD, February 3, 2010). The burn pit operated 24 hours per day, 7 days a week (Taylor et al. 2008).

The committee requested but did not obtain documentation from the DoD on the volumes and types of materials burned specifically at JBB. Although the total quantity of waste burned daily is unknown, the DoD estimates that JBB, with a large population that sometimes surpassed 25,000—including U.S. troops, host nation soldiers, coalition troops, civilians, and contractors— burned as much as several hundred tons a day of waste in the spring of 2007 (Taylor et al. 2008). A subsequent DoD report estimates that the waste stream in spring 2007 was as much as 200 tons per day (USAPHC 2010). By fall 2007, when two incinerators were operational, about half the spring 2007 volume was being burned and by May or June 2009, when three incinerators were operating, only about 10 tons of waste were burned in the pit each day (USAPHC 2010). The burn pit ceased operating in late 2009.

DoD studies and fact sheets prepared for service members stationed at JBB indicated that the pit likely burned a heterogeneous mixture of food waste (including food items, styrofoam, and other related materials), human waste, shipping and packaging materials, meals-ready-to-eat packages, chemicals (paints, solvents), metal/aluminum cans, petroleum, and jet fuel, which was used as and accelerant (Taylor et al. 2008; CHPPM undated). Electronics, tires, batteries, and clothing are not listed specifically, but it is plausible they were burned as well. Based on congressional testimony by individuals who had been present at JBB, medical waste (including needles, gloves, bandages, body fluids, and expired pharmaceuticals) was also burned at least occasionally in the pit (U.S. Congress 2009a).

MILITARY BURN PITS POLICIES AND STUDIES

Although open-air waste burning has long been used by troops in combat situations, concerns regarding the possible health effects and environmental impacts created by such burning have only recently been addressed. Combat situations pose so many other grave risks that the negative aspects of military waste burning have historically been largely ignored. Concerns about possible health risks associated with smoke from open-air waste burning can be traced back in part to the 1990–1991 Persian Gulf War. In response to a constellation of unexplained symptoms and illnesses reported by returning Gulf War veterans, the DoD, the Department of Veterans Affairs (VA), and Congress sponsored a series of studies. These studies indicated that exposures to smoke from oil-well fires and from other combustion sources, including waste burning, were stressors for troops serving in the Gulf War (IOM 2005).

Peace-keeping operations in the Balkans in the 1990s offered the U.S. military an opportunity to examine patterns of waste management and to establish waste management guidelines and policies to reduce health and environmental impacts. During Operation Joint Endeavor in Bosnia in 1995–1996, military preventive-medicine personnel recognized that open burning of waste might be an operational necessity during combat operations, but they emphasized that burning should be used to the minimum extent feasible and that burn pits should be located as far as possible downwind of personnel (U.S. Congress 2009b). Open-air waste burning in Bosnia and Kosovo was replaced by other waste management practices, including incinerators.

A comparative study of waste generation and management practices at several U.S. bases in the Balkans found that bases differed in the generation of plastic waste, primarily from drinking water bottles. The waste stream at one base was comparable to a similarly sized civilian community except for an "extraordinarily large volume of plastic water bottles," which was a cause of concern in part because of the health risks associated with burning plastics (U.S. Army Corps of Engineers 2003).

When military operations began in Afghanistan in 2001 and in Iraq in 2003, there were no regulations governing the use of burn pits in overseas combat operations. An Overseas Environmental Baseline Guidance Document (DoD 2007) did offer guidance on their appropriate use as part of a waste management program, but this document excluded "contingency operations" and thus was not enforceable in Operation Enduring Freedom (OEF) in Afghanistan and Operation Iraqi Freedom (OIF) in Iraq, both of which were defined as such. In the absence of formal policy, operation orders were issued, until 2007 when the DoD issued a policy that included contingency operations; revised regulations were released in 2009 (CENTCOM 2009).

The 2008 study by the U.S. Army Center for Health Promotion and Preventive Medicine (CHPPM) at JBB was not the first DoD study of the possible health risks associated with waste burning. A series of risk assessment studies of a waste incinerator near the U.S. Naval Air Facility in Atsugi, Japan, were conducted by the Navy Environmental Health Center beginning in 1995 in response to local concerns about breathing toxic chemicals (NRC 2001). In 2001, the CHPPM performed a health risk assessment of the municipal and medical waste incinerators at Camp Bondsteel in Kosovo (CHPPM 2001) identifying coarse particulate matter (PM) exposure as a moderate threat at one location (but with low confidence) and providing recommendations for improved operation of the incinerators. A follow-up study in 2005 found that all exposures had low hazard severity and probability (CHPPM 2005).

A CHPPM study of possible health risks specifically from waste burning was conducted in 2004 at Camp Lemonier in Djibouti. There was concern over health risks posed to troops from exposure to smoke from an off-base burn pit located approximately two kilometers from the camp. The burn pit was administered by local civilians, who burned items from both the U.S. base and from the local community. The study concluded that there was a "moderate" overall risk, primarily due to levels of acrolein, aluminum, and PM_{10}. The report recommended that burn barrels be used instead of burn pits and that exposure of military personnel be reduced by putting limits on access to the base and restrictions on outdoor activity during periods of heavy smoke (CHPPM 2004).

The impacts of burn pits were also considered in a 2008 RAND report on environmental concerns for overseas combat operations. The report, commissioned by the U.S. Army, considered burn pits as just one among many case studies, and concluded that "environmental considerations are not well incorporated into Army planning or operations." The report argued that while Army leadership has not always seen environmental concerns as a priority in contingency operations, they should, for two main reasons: "First the environment can affect the health and safety of soldiers. Second, the environment can affect the ability of commanders to accomplish their mission and achieve U.S. national objectives." To change how environmental issues are dealt with in contingency operations, the report urged changes in policy and in the message from commanders who must promote the belief that environmental issues are critical to military missions. The attitude shared by commanders and troops that environmental considerations were not important contributed to the dumping of waste without taking basic precautions with potentially hazardous waste. The report discussed the policy implications of labeling bases "temporary," even after years of habitation, because it limited available funding for alternative waste management options such as incinerators (RAND 2008).

FEDERAL GOVERNMENT'S RESPONSE TO CONCERNS REGARDING BURN PITS

Department of Defense Studies

In response to personnel complaints of odor, poor visibility, and health effects (eye and respiratory irritation) attributed to burn pit emissions, the CHPPM and the Air Force Institute for Operational Health (AFIOH) conducted ambient air sampling and a screening health risk assessment in the spring of 2007. The assessment was designed to detect potentially harmful inhalation exposures for personnel at JBB from chemicals expected to be released from the burn pit (Taylor et al. 2008). A review of the screening health risk assessment conducted by the Defense Health Board (DHB 2008) was answered by an Addendum to the initial report, and a second Addendum followed further sampling in the fall of 2007 (CHPPM and AFIOH 2009). Follow-up sampling in 2009, after most of the waste was diverted to incinerators led to another, similar assessment (USAPHC 2010).

Air samples were initially taken between January and April 2007 when an estimated 200 tons of waste per day were being burned in the pit. Based on those air sampling results, a screening health risk assessment was conducted to estimate potential cancer and noncancer risks to personnel serving at JBB. The CHPPM report indicated that the risk of acute health effects from all substances detected except PM_{10} was low, with screening-level cumulative hazard indices below 1.0 and acceptably low cancer risk estimates for exposures of up to 15-months duration. For noncancer endpoints, the hazard quotient is the ratio of a chronic daily intake for a specific chemical to the toxicological reference dose for that chemical. For JBB this would be the inhalation reference dose. A sum of the hazard quotients for all chemicals at a site is the hazard index. A hazard index of less than 1.0 is considered "safe" or "acceptable" by EPA (EPA 2000). For cancer endpoints, EPA guidance indicates that a risk ranging from 1 in 10,000 to 1 in 1,000,000 or lower is "safe" or "acceptable" (EPA 2000). The CHPPM made several recommendations for engineering controls and better planning to reduce exposure to burn pit emissions, as well as recommending further study, improved risk communication, and policy review. See Chapters 4 and 5 for more details on the measurements and assessments in the CHPPM study.

The second addendum to the screening health risk assessment included an analysis of additional air samples collected from October through November 2007 at JBB, after two incinerators were operational, diverting about half the waste previously going to the burn pits. The fall 2007 sampling data were used for a human health risk assessment that produced results similar to the earlier study. Risks for both cancer and noncancer health outcomes were considered to be "acceptable" for those stationed on the base for up to 15 months; again except for PM_{10} all measured air concentrations were within 1-year military exposure guidelines. However, the results of the risk assessment indicated screening-level cumulative hazard indexes greater than unity for some of the exposure locations and time periods, although cancer risk estimates were still acceptably low. Similar risks were estimated using the May–June 2009 sampling data, when almost all the waste was being burned in the on-site incinerators and only about 10 tons of waste per day was being burned in the pit (USAPHC 2010).

Congressional Responses

After the release of the initial screening health risk assessment in May 2008, which found no cancer or non-cancer health risks attributable to exposure to burn pit emissions at JBB, Congress held a series of public hearings, and several bills on the subject were introduced. Hearings during the fall of 2009 included testimony from both military officials and veterans groups. Much of the testimony focused on the CHPPM screening report, with DoD and VA officials emphasizing the study's conclusion that exposures at JBB fell within DoD and EPA values for acceptable risk. Military officials outlined efforts to enlarge and improve surveillance and assessment activities by Army and Air Force medical and environmental units, as well as enhanced coordination with the VA on the surveillance and treatment of any possibly related medical issues.

Other witnesses presented first-hand accounts of the burn pits. During testimony and in media interviews, veterans and individual medical and environmental professionals attributed a range of medical problems to smoke from burn pits, including asthma, joint pain, cancer, vomiting and nausea, burning lungs, and Parkinson's disease (U.S. Congress 2009a). Witnesses testified to the pervasive presence of noxious smoke in their tents and emphasized the limitations and shortcomings of the CHPPM report. Medical and environmental personnel reported increased respiratory symptoms among service members in Iraq and in returning veterans seeking medical treatment stateside.

In October 2009, the National Defense Authorization Act for FY 2010 was passed. The law included language (Section 317) prohibiting the disposal of waste in open-air burn pits by the DoD and called for the DoD to issue appropriate regulations.

In 2009, Congress also requested that the GAO investigate the use of burn pits in Iraq and Afghanistan; the guidance available on burn pits and adherence to that guidance; alternatives to burn pits and the DoD's implementation of these alternatives; and the efforts made by the DoD to evaluate air quality and exposures in accordance with applicable guidance. The GAO report, released in October 2010, recommended that the DoD issue, implement, and ensure adherence to guidance for burn pit operations and waste management; conduct monitoring of burn pits as directed by guidance; characterize the waste produced and implement waste management strategies to reduce hazards from burning waste; and consider alternatives to burn pits (taking into account costs, feasibility, and health effects) (GAO 2010).

Department of Veterans Affairs Responses

In 2009, the VA requested that the Institute of Medicine (IOM) convene a committee to determine the potential long-term health effects of exposure to burn pits in Iraq and Afghanistan, and to specifically examine the burn pit at JBB in Iraq. That request resulted in this study.

The VA also issued a training letter in April 2010 to provide guidance to regional offices on how to handle claims for disabilities related to specific environmental hazards, including burn pits. Based on the 2008 CHPPM report (Taylor et al. 2008), the VA recognized that exposure to polycyclic aromatic hydrocarbons, volatile organic compounds, dioxins and furans, and PM may result from the use of burn pits, but that potential health effects had not yet been determined. Owing to the widespread use of burn pits and the inability of the DoD to identify all duty locations for all personnel, a veteran's "lay statement" is sufficient to establish exposure to burn pits if they served in Iraq, Afghanistan, or Djibouti. The VA noted that rating authorities should be prepared to review claims and recognize the potential for burn pit exposures because veterans suffering from respiratory, cardiopulmonary, neurological, autoimmune, or dermal disorders may not associate these illnesses with exposure to burn pits. The training letter also identified PM from diverse local sources as a specific environmental hazard in Iraq, Afghanistan, and Djibouti, and thus regional VA offices were instructed to be prepared to review claims for respiratory and cardiopulmonary health effects (both have been associated with exposure to PM) that may be associated with exposure to PM (VA 2010).

CONCLUSIONS

Although the practice of open-air waste burning has a long history in the U.S. military, a combination of factors, including the large volume of waste burned, the increase in the waste stream of items such as plastics with the potential to produce toxic materials when burned, and increased public awareness of health risks associated with exposure to air pollution, has led to greater concern among service members and their families about possible long-term health effects resulting from exposure to burn pit emissions at military bases in Iraq and Afghanistan. Given the nature of military conflicts and the constant movement of troops and supplies, it is difficult to characterize waste disposal practices at many bases, or apply knowledge of one base to other bases. Although some monitoring of air pollution has been conducted at JBB and preliminary screening risk assessments developed, the air quality and sources of pollution at JBB and other bases are still in question, leading to further concerns about the possible health risks that may ensue from exposure to these pollutants. Although several government agencies are involved in examining burn pits, issues related to their use and safety are unresolved. To help address some of these issues, the VA requested that the IOM conduct this study.

REFERENCES

CENTCOM (U.S. Central Command). 2009. *Environmental quality CENTCOM contingency environmental guidance*. Document number R 200-2. MacDill Air Force Base, FL: Headquarters, U.S. Central Command.

CHPPM (U.S. Army Center for Health Promotion and Preventative Medicine). 2001. *Environmental surveillance health risk assessment no. 47-EM-2638-01D. Environmental surveillance of Camp Bondsteel incinerators 18–28 January 2001*. With redaction of author information. Camp Bondsteel, Kosovo: U.S. Army Center for Health Promotion and Preventive Medicine.

CHPPM. 2004. *Ambient air assessment for Camp Lemonier, Djibouti, 4–10 Jun 2004*. Aberdeen Proving Ground, MD: U.S. Army Center for Health Promotion and Preventive Medicine.

CHPPM. 2005. *Transmittal of results for deployment occupational and environmental health (OEH) risk characterization*. With redaction of author information. Aberdeen Proving Ground, MD: U.S. Army Center for Health Promotion and Preventive Medicine.

CHPPM. Undated. *Just the facts: Balad burn pit*. Aberdeen Proving Ground, MD: U.S. Army Center for Health Promotion and Preventive Medicine.

CHPPM and AFIOH (U.S. Air Force Institute for Operational Health). 2009. *Addendum 2. Screening health risk assessment burn pit exposures, Balad Air Base, Iraq, May 2008*. USACHPPM Report No. 47-MA-08PV-08/AFIOH Report No. IOH-RS-BR-TR-2008-0001. Aberdeen Proving Ground, MD: U.S. Army Center for Health Promotion and Preventive Medicine. August.

DHB (Defense Health Board). 2008. *Defense Health Board findings pertaining to health risk assessment, burn pit exposures, Balad Air Base, Iraq. June 26*. Falls Church, VA: Defense Health Board.

DoD (U.S. Department of Defense). 2007. *Overseas environmental baseline guidance document*. DoD 4715.05-G. Washington, DC: Office of the Under Secretary of Defense for Acquisitions, Technology, and Logistics.

DoD. 2011. *Exposure to toxins produced by burn pits: congressional data request and studies*. In Memorandum for the assistant secretary of defense for health affairs. Washington, DC: U.S. Department of Defense.

EPA (U.S. Environmental Protection Agency). 2000. *Science Policy Council handbook: risk characterization*. EPA-100-B-00-002. Washington, DC: Office of Research and Developoment.

EPA. 2011. *Backyard burning*. http://www.epa.gov/epawaste/nonhaz/municipal/backyard/index.htm (accessed March 2, 2011).

GAO (U.S. Government Accountability Office). 2010. *Afghanistan and Iraq: DOD should improve adherence to its guidance on open burning and solid waste management*. GAO-11-63. Washington, DC: U.S. Government Accountability Office.

IOM (Institute of Medicine). 2005. *Gulf War and health: Volume 3. Fuels, combustion products, and propellants*. Washington, DC: The National Academies Press.

Melosi, M. V. 2005. *Garbage in the cities: Refuse, reform, and the environment*. Pittsburgh, PA: University of Pittsburgh Press.

NRC (National Research Council). 2001. *Review of the U.S. Navy's human health risk assessment of the Naval Air facility at Atsugi, Japan*. Washington, DC: National Academy Press.

RAND. 2008. *Green warriors: Army environmental considerations for contingency operations from planning through post-conflict*. Arlington, VA: RAND Corporation.

Taylor, G., V. Rush, A. Deck, and J. A. Vietas. 2008. *Screening health risk assessment burn pit exposures Balad Air Base, Iraq, and addendum report*. IOH-RS-BR-TR-2008-0001/USACHPPM 47-MA-08PV-08. Brooks City-Base, TX: Air Force Institute for Operational Health and U.S. Army Center for Health Promotion and Preventative Medicine.

U.S. Army Corps of Engineers. 2003. *Analysis of the waste management practices at Bosnia and Kosovo base camps*. ERDC/CRREL TR-03-6. Hanover, NH: U.S. Army Corps of Engineers.

U.S. Congress, Senate, Democratic Policy Committee. 2009a. *Are burn pits in Iraq and Afghanistan making our soldiers sick?* 111th Cong., 1st Sess. November 6.

U.S. Congress, Senate, Committee on Veterans' Affairs. 2009b. *VA/DOD response to certain military exposures*. 111th Cong., 1st Sess. October 8.

USAPHC (U.S. Army Public Health Command). 2010. *Screening health risk assessments, Joint Base Balad, Iraq, 11 May–19 June 2009*. With redaction of author information. Aberdeen Proving Ground, MD: U.S. Army Public Health Command. July.

VA (Department of Veterans Affairs). 2010. *VA environmental hazards memo*. 211A training letter 10-03. Washington, DC: U.S. Department of Veterans Affairs.

Weese, C. B., and J. H. Abraham. 2009. Potential health implications associated with particulate matter exposure in deployed settings in southwest Asia. *Inhalation Toxicology* 21(4):291-296.

3

Approach to the Task

The organizing principles, themes and structure for this study are derived from two sources. First, the statement of task asks the Institute of Medicine (IOM) to determine "the potential long-term health effects of exposure to burn pits in Iraq and Afghanistan" and to design a future study of long-term health effects in veterans who may have experienced these exposures. Although the statement of task specifies that Joint Base Balad (JBB) in Iraq be considered as an example of a military base with a burn pit, it is clear that the Department of Veterans Affairs (VA's) interest extends to all military personnel deployed to bases in Iraq and Afghanistan who may have been exposed to burn pit emissions. These dual objectives shaped the committee's comprehensive search for information on relevant exposures and health outcomes, as well as its highly focused analyses of data and information from military reports specific to JBB and other burn pit locations.

Second, meeting the statement of task objectives entailed an in-depth examination of a substantial body of information on exposure to combustion products and health effects. For example, the health effects literature on dioxin-like compounds and particulate matter (PM) is vast, and they are but two of the numerous and diverse environmental agents likely to be found in burn bit emissions. Similarly, a cursory review of the scientific literature reveals a wide range of health outcomes—cancer, pulmonary disease, cardiovascular disease, neurological effects—in human populations exposed to environmental agents found in combustion products, singly and in mixtures. Comparable studies on laboratory animals often complement and supplement the data from human studies.

Effective use of this wide-ranging literature requires a systematic approach. The "risk assessment paradigm" introduced in the seminal report *Risk Assessment in the Federal Government, Managing the Process* (NRC 1983) offers a simple, flexible template. Drawing on advances in the theory and practice of risk assessment in the regulatory context, subsequent reports and guidelines have modified and expanded the paradigm to take into account issues specific to, for example, air pollutants (NRC 1994), ecological risk (EPA 1998), and the relationship between risk assessment and decision-making (NRC 2009). This paradigm, now called a conceptual framework, has been updated with more emphasis on problem formulation and improvements in technologic analyses and to provide more guidance on expanding the utility of risk assessments (NRC 2009). As shown in Figure 3-1, this paradigm calls for several analyses:

- Exposure assessment—including the analysis of exposures, qualitative and quantitative, currently experienced or anticipated under existing or expected conditions;

- Hazard identification—which requires determining whether exposure to a particular environmental agent has the potential to increase the incidence of an adverse health effect and, for particular exposure conditions, which agents can lead to adverse effects, thus requiring follow-up through dose–response or exposure–response analysis; and
- Exposure–response/dose–response or toxicity assessment—which quantifies the relationship between the exposure or dose in humans and the incidence of adverse effects or health outcomes in human populations or laboratory animals; and Risk characterization—which synthesizes and integrates data and information to develop a quantitative estimate of the incidence of the effect in a given population, along with a discussion of related qualitative considerations, strengths, limitations, and uncertainties.

This report is organized around the assessment process with modifications to address the specific issues raised in the statement of task. As shown in Figure 3-2, the process focuses first on research and data collection related to exposures at JBB, health effects that might result from those exposures, and health outcomes in other human populations exposed to some of the environmental agents found in ambient samples and potentially emitted by the JBB burn pit. Based on this information, assessments are then made on potential exposures and health effects that might occur in relevant populations. Rather than conduct a formal, quantitative risk assessment and risk characterization, the committee prepared a synthesis and summary of key findings and applied that informa-

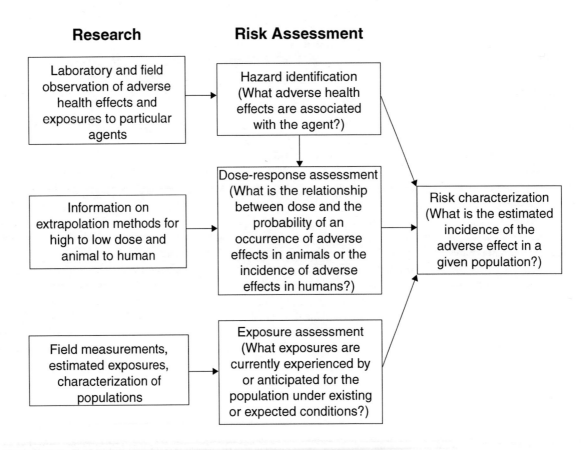

FIGURE 3-1 Elements of the risk assessment paradigm.
SOURCE: Adapted from NRC (1983, 2009).

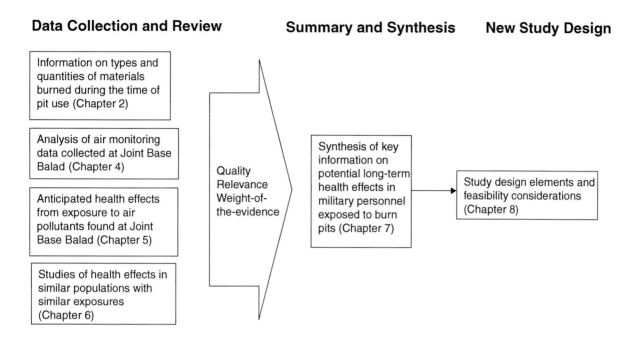

FIGURE 3-2 Approach to the statement of task: The types of information reviewed and the process used to develop this report.

tion to suggest design elements for a future epidemiologic study to address data gaps. The resulting three-stage analytical process is outlined below:

1. Data collection and review: For this report, the "exposure assessment," "hazard identification," and "dose–response" tasks in the 2009 framework are merged as "data collection and review" activities focused on chemicals detected in air monitoring at JBB and on smoke exposures per se. These activities revolve around two closely related but distinct topics: exposure and health effects. The in-depth review begins with identifying and selecting agents and corresponding reliable studies to be given priority for evaluating health outcomes potentially associated with burn pit exposures. This includes establishing criteria for evaluating the quality and relevance of the studies along with their limitations and uncertainties (see Chapters 4, 5, and 6).
2. Synthesis: Assessing "the potential long-term health effects of exposure to burn pits in Iraq and Afghanistan" requires the identification and evaluation of the key findings from the comprehensive review of environmental monitoring data and health effects information. One focus is on identifying which, if any, of the many environmental agents detected in the air at JBB are present in toxicologically significant quantities, and which, if any, are associated with the burn pit, and determining the reliability of those data. Another focus is the major findings on health effects seen in other similarly exposed populations as discussed in Chapter 6, including the extent to which these populations experienced exposures relevant (with respect to chemical composition and level of exposure) to those at JBB. A related question is how comparable those populations are to military personnel stationed at JBB. This synthesis includes a discussion of the committee's confidence in the quality and utility of these findings for the VA (see Chapter 7).
3. New study design: The exposure–response information developed in Chapters 5 and 6, combined with information on the potential exposures of JBB personnel discussed in Chapter 4, provide the basis for evaluating feasibility and design issues specific to an epidemiologic study to assess health outcomes associated with exposure to burn pit emissions at JBB and other burn pit sites (see Chapter 8).

Each of the following sections corresponds to one of the three steps in the analytical process. The next section, "Data Collection and Review," outlines the kinds of data and information needed to evaluate long-term health effects in line with the statement of task and related risk assessment principles, and explains the principles and processes guiding the evaluation and selection of data presented in the remainder of the report. In the section titled "Summary and Synthesis," the committee summarizes and highlights critical findings from the previous section. In the final section, "New Study Design," this information is used to propose design elements for a future epidemiologic study of health effects in veterans who may have been exposed to burn pit emissions at JBB.

DATA COLLECTION AND REVIEW

Data collection for this report centered on two general topics: first, information on environmental releases and concentrations of combustion products at JBB and the resulting potential human exposures, and second, the potential for long-term health effects resulting from those exposures. Principles and methods guiding the committee's assessment of the data are outlined in the following sections.

Environmental Exposure

Exposure assessment requires characterization of the frequency, magnitude, and duration of exposure to an agent of concern in an exposed population (NRC 2010). The committee began assessing possible burn pit emission exposures by reviewing Department of Defense (DoD) air monitoring data collected at JBB and attempting to determine whether the air pollutants detected during the sampling periods could be attributed to the burn pits or might possibly originate from other sources, either natural or anthropogenic (see Chapter 4). Three DoD documents were used to characterize ambient concentrations: the 2008 *Screening Health Risk Assessment Burn Pit Exposures, Balad Air Base, Iraq and Addendum Report* (Taylor et al. 2008), a second addendum to the report (CHPPM and AFIOH 2009), and a follow-up *Screening Health Risk Assessments* study in 2010 (USAPHC 2010). The committee asked for and received from DoD raw air-sampling data collected in 2007 and 2009 to help in identifying potential sources of the air pollutants. The raw data was not only useful for determining which chemicals were tested for in the samples, but they were also useful in determining which chemicals were found at concentrations above the analytical detection limit. Those chemicals present above the detection limit were considered to warrant further assessment.

The committee recognized that air quality in a region is affected by many factors, such as temperature, humidity, meteorological events, and atmospheric inputs from both natural sources such as dust storms and from anthropogenic sources such as power plants, agriculture, and industrial facilities. Thus, the committee asked for and received from DoD meteorological information for the JBB area during the air monitoring periods.

The location and activities of the military personnel stationed at JBB was also a factor for assessing exposure. The committee noted that not only did the personnel work on the base, but they also lived there and thus were potentially exposed to the burn pit emissions 24 hours a day, 7 days a week, although potentially at different levels during on- and off-work hours. Furthermore, it was assumed that some personnel would have greater exposure to burn pits either because they actually worked in the pits or may have had a job that required them to work near the pit. The committee did not ask the DoD for specific information regarding the number of military personnel that worked at or in close proximity to the burn pit. Because JBB had both Army and Air Force personnel, and the base was used as a transit stop for some personnel, the duration of exposure to the burn pit emissions was also highly variable. The committee assumed that some unknown number of personnel would experience high exposure to burn pit emissions based on their military occupation specialty and their housing location and other personnel might have relatively low exposure.

Potential Long-Term Health Effects

Assessing potential long-term health effects from exposure to environmental agents depends on two closely related but different aspects of exposure. The *environmental exposure* information developed in Chapter 4 describes the occurrence and concentration of environmental agents in external media, in this case air. This information can provide a starting point for evaluating internal exposure—the presence and behavior of agents in living systems—and the biological response to that exposure. Alternatively, the information may be used directly in exposure-response evaluations without a full evaluation of dose. Ideally, however, assessment of health effects includes data on internal dose—the amount of agent in the tissues of the organism. Many factors determine internal dose, including the concentration of the agent in environmental media, the physical and chemical properties of the agent, and individual metabolic processes of the organism. Information on the majority of these exposure factors for military personnel living and working at JBB is unavailable or incomplete. Thus, the committee conducted its review assuming that deployed personnel were exposed to burn pit emissions primarily through inhalation and they received some unquantified internal dose of the chemicals present in burn pits emissions. The committee recognized that some dermal and ingestion exposure was possible, albeit at relatively minor levels. However, the lack of data on contaminant levels in soil, food, water, and surfaces at JBB precluded further consideration of ingestion and dermal routes of exposure.

Studies on Health Effects

For chemical mixtures such as burn pit emissions, toxicity and other health effects data on the mixture *itself* are generally scarce or nonexistent. In view of this paucity of data, the committee focused on studies of exposure to individual chemicals, and exposure to similar mixtures in surrogate populations. First, studies of exposure to known *individual* chemicals in laboratory animals provide information on health effects (or lack thereof) attributable to specific chemicals at relevant exposure levels (Chapter 5). These studies can be important sources of both *qualitative* (the nature and range of health effects observed such as cancer, neurological effects, and reproductive effects) and *quantitative* (dose–response or exposure–response relationships) information on health effects in test animals associated with exposure to specific chemicals under controlled laboratory conditions. However, the extent to which animal studies are predictive of health effects (or lack of effects) in humans varies considerably.

The variety of materials burned at JBB produced a complex mixture of chemicals of differing toxic potential, occurring in different amounts and physical forms. The emissions from the burn pits were also mixed with numerous air pollutants from other local and regional sources such as industry and dust storms. Based on the air monitoring data analyzed in Chapter 4, the committee determined the likely adverse health effects that might be associated with those chemicals found above the analytical detection limit or that otherwise were expected to pose the greatest risk to JBB personnel. To identify potential health effects from these exposures, the committee relied on published summaries from diverse sources, including IOM and NRC reports, government reports such as toxicological profiles from the Agency for Toxic Substances and Disease Registry, and established databases such as the U.S. Environmental Protection Agency's (EPA's) Integrated Risk Information System (IRIS).[1] The committee did not re-examine the underlying data or methodology for these sources; rather, it relied on them as well established sources of health effects information (see Chapter 5).

Epidemiologic Studies

Given the lack of information on health effects that may result from exposure to burn pit emissions per se, the committee sought information on health effects from epidemiologic studies of other (non-Balad) human populations exposed to smoke, combustion products, or other complex chemical mixtures that include at least some of the chemical constituents present in burn pit emissions (see Chapter 6). The committee gave special attention to studies involving other military populations, firefighters, and residents living near municipal waste incinerators. Those studies provide data on demographically similar populations (relatively young and healthy, and predomi-

---

[1] Although scientists throughout the world use the IRIS database for its valuable toxicity and risk information, many IRIS profiles need to be updated to take into account new pollutant-specific data and advances in methodology or to correct inaccurate values (NRC 2008). Although the committee recognizes these concerns and the resulting uncertainties they create, these values, with those from other health agencies, are the best readily available data for assessing pollutants emitted from open burning.

nantly male) exposed to similar pollutants (chemical mixtures produced by burning). In selecting studies for this report, the committee focused on factors such as exposure duration, populations of interest, criteria for identifying key studies, and determining categories of association between exposures and health effects.

Exposure Duration

In general, the available studies involved one of three exposure–response patterns: (1) medium-term exposure (6 months to a year) to higher pollutant levels that might increase the risk of chronic disease later in life (medium-term exposure/long-term health effects); (2) short-term exposure (over a few days) to higher pollutant levels that might increase the risk of chronic disease later in life (acute exposure/long-term health effects); and (3) short-term exposure to higher pollutant levels that might increase the risk of an acute adverse health event immediately following or shortly after exposure (acute exposure/acute health effects). For relevance to veterans of the current conflicts, the committee focused on long-term health outcomes that persist after exposure ceases, rather than acute health effects.

Populations of Interest

Assessing potential future effects of exposure in a particular population such as JBB personnel depends on available information for that population together with observational data on effects in theoretically similar populations exposed under generally comparable conditions. Three occupational populations were of special interest to the committee because of their exposure to complex mixtures of combustion products potentially similar to burn pit emissions: firefighters (wildlands and urban), municipal waste incinerator workers, and military personnel from several earlier deployments (Bosnia, Kuwait, Iraq, Afghanistan). The IOM has evaluated the scientific literature on military personnel deployed to the 1990–1991 Persian Gulf War in previous reports (IOM 2005, 2006, 2010), which this report incorporates by reference where appropriate. Residents living in the vicinity of municipal waste incinerators were also included in the committee's deliberations as personnel at JBB not only work on the base but live there as well with the potential for 24-hour exposure. Finally, the committee sought, but did not find, epidemiologic information on health effects seen in Iraqi civilians living near bases with burn pits or other sources of combustion products, such as local waste burning operations or wild fires.

Selection of Key Studies

The studies reviewed in Chapter 6 are designated "key" or "supporting" depending on several criteria as described in that chapter. Consistent with previous IOM reports (IOM 2010), to be designated "key" a study had to include information about the putative exposure and specific health outcomes, demonstrate rigorous methods, include methodological details sufficient to allow a thorough assessment, and include an appropriate control or reference group. A "supporting" study typically had methodologic limitations, such as lack of a rigorous or well-defined diagnostic method or lack of an appropriate control group. The committee's evaluations of key and supporting studies used in reaching its conclusions are described in greater detail in Chapter 6.

Categories of Association

The committee considered the studies identified in Chapter 6 in terms of the strength of the apparent association between the exposures to combustion products and potential health effects. As detailed in Chapter 6, the five categories used by the committee represent different strengths of association and different levels of confidence in the potential relevance and utility of each study. The validity of an observed association is likely to vary with the extent to which common sources of spurious associations can be ruled out as contributing factors. Accordingly, the criteria for each category express a degree of confidence based on the extent to which sources of error were reduced.

SUMMARY AND SYNTHESIS

Understanding the potential for long-term health effects associated with human exposure to burn pit emissions depends on the reliability and utility of the available data on exposures at JBB and other burn pit locations, as well as exposure and outcome data from studies on other populations exposed to some of the same combustion products, singly or in mixtures. Each of the three main data sources—data specific to JBB, other military reports, and the peer-reviewed literature—provides useful and relevant information, but each has major limitations.

In this section, the committee summarizes and highlights key findings on materials burned at JBB and other military burn pit locations, health effects data on the combustion products detected at JBB, and studies on health effects in non-Balad populations potentially exposed to similar chemicals. The committee comments on its confidence in these findings and on their usefulness in providing the VA with information for medical follow-up and future studies. The committee also considers the possible impact of co-exposure to combustion products from local and regional air pollution sources other than the JBB burn pit.

NEW STUDY DESIGN

The statement of task directs the committee to "examine the feasibility and design issues for an epidemiologic study of veterans exposed at the Balad burn pit." Two recent IOM reports on other military populations offer useful models. First, a study on depleted uranium in military and veteran populations (IOM 2008) identifies key features of a well-designed epidemiology study of potential health effects associated with environmental exposure.

Second, the review of the DoD's Enhanced Particulate Matter Surveillance Program (NRC 2010) identifies exposure assessment as a central element in the study design for that proposed study because "if data are not available to characterize exposure, any study of association between health outcomes and exposure will not provide valid results." Any study of JBB exposures is necessarily incomplete because of the limited exposure information available. Specifically, the military collected *environmental* samples at fixed locations, but the relation to *individual* exposures may be difficult to ascertain, be subject to large uncertainties, or both. Since the JBB burn pit was closed in 2009, no additional opportunities are available to collect environmental or personal monitoring data on burn pit emissions at JBB, although it is still possible to conduct air monitoring, including personal monitoring, at other military sites with operational burn pits.

Epidemiologic studies conducted at JBB and other bases with burn pits have substantial acknowledged limitations (for example, lack of adequate follow-up for the diagnosis of conditions with long latency such as chronic bronchitis, emphysema, systemic lupus erythematosus, and cancer). Taking these considerations and the key findings in this report into account, the committee closes with a discussion of design issues for a future epidemiologic study of the JBB population and outlines a proposed approach for that study.

REFERENCES

CHPPM (U.S. Army Center for Health Promotion and Preventive Medicine) and AFIOH (U.S. Air Force Institute for Operational Health). 2009. *Addendum 2. Screening health risk assessment burn pit exposures, Balad Air Base, Iraq, May 2008*. USACHPPM Report No. 47-MA-08PV-08/AFIOH Report No. IOH-RS-BR-TR-2008-0001. Aberdeen Proving Ground, MD: U.S. Army Center for Health Promotion and Preventive Medicine. August.

EPA (U.S. Environmental Protection Agency). 1998. *Guidelines for ecological risk assessment*. EPA/630/R-95/002F. Washington, DC: Risk Assessment Forum.

IOM (Institute of Medicine). 2005. *Gulf War and health: Volume 3. Fuels, combustion products, and propellants*. Washington, DC: The National Academies Press.

IOM. 2006. *Gulf War and health: Volume 4. Health effects of serving in the Gulf War*. Washington, DC: The National Academies Press.

IOM. 2008. *Epidemiological studies of veterans exposed to depleted uranium: Feasability and design issues*. Washington, DC: The National Academies Press.

IOM. 2010. *Gulf War and health: Volume 8. Update of health effects of serving in the Gulf War*. Washington, DC: The National Academies Press.

NRC (National Research Council). 1983. *Risk assessment in the federal government: Managing the process*. Washington, DC: National Academy Press.

NRC. 1994. *Science and judgment in risk assessment*. Washington, DC: National Academy Press.

NRC. 2009. *Science and decisions: Advancing risk assessment*. Washington, DC: The National Academies Press.

NRC. 2010. *Review of the Department of Defense Enhanced Particulate Matter Surveillance Program report*. Washington, DC: National Academies Press.

Taylor, G., V. Rush, A. Deck, and J. A. Vietas. 2008. *Screening health risk assessment burn pit exposures, Balad Air Base, Iraq, and addendum report*. IOH-RS-BR-TR-2008-0001/USACHPPM 47-MA-08PV-08. Brooks City-Base, TX: Air Force Institute for Operational Health and U.S. Army Center for Health Promotion and Preventative Medicine.

USAPHC (U.S. Army Public Health Command). 2009. *Screening health risk assessments, Joint Base Balad, Iraq, 11 May–19 June 2009*. With author information redacted. Aberdeen Proving Ground, MD: U.S. Army Public Health Command. July.

4

Evaluation of Air Monitoring Data and Determinants of Exposure

The Department of Defense (DoD) has conducted air monitoring studies at Joint Base Balad (JBB) in Iraq in response to complaints by military personnel stationed there that smoke from the burn pit was causing health problems. A recent National Research Council (NRC) report (2010) reviewed the DoD's Enhanced Particulate Matter Surveillance Program (EPMSP) conducted at U.S. air bases in the Middle East (including JBB) for the U.S. Army's Center for Health Promotion and Preventive Medicine (CHPPM, now the U.S. Army Public Health Command) (Engelbrecht 2008; Engelbrecht et al. 2009). The NRC report found several limitations in the methodology and study design for the CHPPM study. In 2007 and 2009, CHPPM conducted a series of monitoring campaigns at JBB to measure the concentration of airborne pollutants at several sites on the base. The measurements were used as inputs for risk assessments for potential cancer and noncancer effects that might result from exposure to burn pit emissions (Taylor et al. 2008; CHPPM and AFIOH 2009; USAPHC 2010). The results of these monitoring campaigns are reviewed and further analyzed in this chapter.

In this chapter, the expected sources and nature of air pollutants found at JBB are described based on the location of the base and its operations. This description is followed by a summary of the results of air monitoring carried out at the base; an explanation of the limitations and strengths of the monitoring are provided in Appendix B. The monitoring data are used to compare the average chemical composition of air pollution at different locations on the base to pollution profiles for other locations around the world.

POLLUTANT SOURCES AT JOINT BASE BALAD

Occupants of JBB were and are exposed to a combination of regionally and locally generated air pollutants. Regional air pollutants originate at a considerable distance (miles to thousands of miles) from the exposure location, and may undergo some atmospheric chemical and physical transformations prior to exposure of a receptor. Those pollutants may come from a combination of sources such as industrial activities, mobile sources, and windblown dust. Sources of locally generated air pollutants at JBB include windblown dust, local combustion sources, and volatile evaporative emissions. Local combustion sources include the burn pit or other incinerators for refuse, compression ignition vehicles, aircraft engines, diesel electric generators, and local industry and households. Volatile evaporative emissions come primarily from refueling and other fuel management activities on the base.

At JBB, the high windblown dust concentrations combined with emissions that are combustion derived or from unique regional and local sources offer an unusual mixture of exposures. The particulate matter (PM; primarily windblown dust) contains large amounts of geological materials (for example, aluminum silicates, calcium carbonate, and iron oxide) from local soils, carbon that originates mostly from combustion sources, metals from soils, and emissions from a combination of local and regional mobile sources, including smelting activities. The presence of metals in previously collected EPMSP PM samples illustrate the potential for smelting activities and lead from gasoline (lead is still used in gasoline in this area) to contribute to metal concentrations in the air near highly populated areas (Engelbrecht et al. 2009). Gaseous pollutants such as sulfur dioxide and carbon monoxide may originate locally, such as from combustion and ignition engine sources, but pollutants such as ozone may originate regionally and be generated primarily during photochemical transformations.

The major local pollutant sources at JBB include (or included) combustion products from a combination of airport traffic (airplanes and helicopters), ground transportation, stationary power generation (diesel electricity generators), local industry and households, and waste burning associated with incineration (currently) or the burn pit (previously). Each of these sources emits a complex mixture of particulate and gaseous pollutants that include volatile organic compounds (VOCs), particle- and vapor-phase semivolatile organics, metals, and PM.

In accordance with DoD Directive 4140.25 (April 2004), aircraft, ground vehicles and power generators at JBB are mostly fueled by JP-8 jet fuel, a heavy petroleum distillate fuel similar in characteristics to commercial fuel oil (NRC 2003). Vehicles on and around the base are typically not equipped with emissions reduction technology. The PM from vehicles is typically 100 nm or smaller in diameter at the exhaust and is composed of a mixture of elemental and organic carbon that varies with the engine operating conditions, together with traces of metal oxides. Atmospheric transport of vehicle particulate emissions leads to larger particle sizes as the small particles agglomerate, and some of the semivolatile organics emitted as vapors will condense on particles.

The gaseous- or vapor-phase emissions from the sources affecting JBB include nitrogen oxides, ozone, carbon monoxide, sulfur dioxide, and volatile and semivolatile organic compounds. As discussed in more detail below, the volatile hydrocarbons measured at JBB include other hazardous air pollutants such as formaldehyde, benzene, and 1,3-butadiene, while the measured semivolatile hydrocarbons include polycyclic aromatic hydrocarbons (PAHs), and polychlorinated dibenzo-p-dioxins and furans (PCDDs/Fs), all measured so as to capture both vapor-phase and particulate-phase components. There is significant overlap in the composition of emissions from the various sources, making source attribution difficult or impossible based on simple characterization of ambient air composition. Nevertheless, the subtle differences in observed composition between locations on the base were used by the committee in an attempt to estimate the contributions of the hypothesized major sources of pollutants.

JOINT BASE BALAD MONITORING DATA

Figure 4-1 is a map of the basic layout of JBB, including the monitoring sites for the CHPPM screening health risk assessments (Taylor et al. 2008; CHPPM and AFIOH 2009; USAPHC 2010), and the wind rose for 2003 through 2007. Air measurements were taken at the five sampling sites labeled as mortar pit, guard tower, transportation field, H-6 housing, and Contingency Aeromedical Staging Facility (CASF). The wind direction is primarily northwest to southeast. The committee requested from the DoD, but did not receive, more precise relative coordinate information for the sampling sites and the burn pit and more information on the dimensions of the burn pit. The committee was informed by the DoD that it was unable to provide further information on the location of the burn pit until U.S. troops had left the area (Major Scott Newkirk, Army Institute of Public Health, personal communication, October 28, 2010).

The first air monitoring campaign was conducted from January to April 2007. At that time no incinerators were operating on the base and an estimated 200 tons of waste were burned daily in the pit (USAPHC 2010). During the October–November 2007 monitoring campaign, the burn pit burn rate was estimated to be half the spring value (100 tons/day) with two incinerators operating, and 10 tons/day during the May–June 2009 monitoring campaign when three incinerators were operating (USAPHC 2010).[1] The incinerators are located at the south end of the site, and emissions from them are not expected to have substantially affected the concentrations measured onsite, at least in

---

[1] The committee was not provided with any information on how these burn rates were estimated.

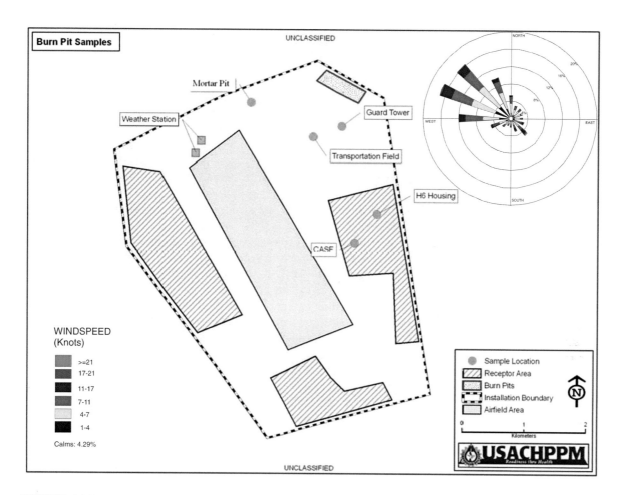

FIGURE 4-1 Sampling points and wind rose at JBB. The length of the vanes in the wind rose corresponds to the fraction of time the wind blows from the direction of the vane. This diagram does not correspond in shape or orientation to the layout found in the aerial photos (from 2004) in Google Earth. Obtaining even approximate correspondence requires a rotation and skewing of this diagram. In particular, the location and size of the burn pit, pointed out to the committee by an Air Force contractor and visible on Google Earth imagery, does not correspond (even after rotation and skewing) to the location or size shown on this diagram. The committee is therefore unclear on the exact locations of the sampling sites relative to the burn pit. SOURCE: Taylor et al. (2008).

2007. [NOTE: There appears to have been sporadic air sampling during 2006; the committee was furnished with some results of 2006 sampling for dioxins but does not know the specific location or exact methodology used.]

The air monitoring approach included fixed site samplers that were placed at the mortar pit (selected to be upwind of the burn pit), at the H-6 housing and CASF sites, and at the guard tower and transportation field sites (chosen to be downwind of the burn pit) (Taylor et al. 2008). No measurements were taken in close proximity to the burn pit. Sampling sites near each other were combined in the 2007 data available to the committee, so that samples were identified as being collected at the mortar pit, the guard tower/transportation field, or H-6 housing/CASF. The 2009 data available to the committee were provided for the five distinct sampling sites shown on Figure 4-1. However, there are no simultaneous samples available to the committee that allow comparison between the guard tower and transportation field sites, or between the H-6 housing and the CASF site. Therefore, the committee combined the data for the guard tower and transportation field sites, and for the H-6 housing and CASF sites, treating each combination of sites as an individual sampling location.

The intended sampling period for all samples was 24 hours, with variation due to the logistics of the sampling operation. No information was available to correlate burn pit operations with measurement times. Individual measurement data were provided in the form of Excel files containing sampling data including sample identifiers, sampling locations (2007 data) or sites (2009 data), sampling times, a few observational field notes, and the individual analytical results for each sample.

Although the samples were tested for a large number of air pollutants, there were a number of air pollutants that were not measured because they were not targeted by the analytical methods used. Notably, this study did not include ozone, carbon monoxide, nitrogen dioxide, or sulfur dioxide, which are criteria pollutants in the United States. Furthermore, many air pollutants that are potentially important in burn pit emissions, such as endotoxins or other biological materials, are rarely monitored and samples were not tested for these substances.

The committee requested and received certain meteorological information from the DoD. Ground-level meteorological measurements are taken regularly at JBB, and CHPPM provided hourly information largely conforming[2] to METAR/SPECI record format (USAF 2009) for JBB for all of 2007 and 2009. Figure 4-1 indicates the location of two weather stations at JBB. The data files suggest a third weather station location on base (the location specified in the file does not correspond with either location shown in Figure 4-1), but none of the meteorological information is particularly location specific. Upper-air sounding data were not available for JBB[3]; however, data were available for Al Asad Air Base, Baghdad; Forward Operating Base Kalsu; Q-West; and Al Taqaddum.

Ambient sampling data are sufficient to draw some important conclusions, for the following reasons. First, the PAH and PCDD/F data in particular are of high quality, as reflected by the high degree of pollutant concentration correlations among sites for near-simultaneous samples. Second, although the small number and timing of samples does not allow determination of representative annual averages, they provide some information about the magnitude of seasonal variability of pollutant concentrations because they were collected during different seasons. Third, most samples were obtained during near-simultaneous sampling of the three locations (at least during the same day), allowing direct comparisons between the sample locations (although as noted in Appendix B, the lack of complete simultaneity does limit the comparability of samples).

The differences in concentrations between sampling locations at JBB for near simultaneous measurements allows evaluation of the effect of local versus background sources of air pollutants, including some inference as to the contribution of the burn pit. In the following discussion, a local source is defined as one that has a differential impact on the three JBB locations, in contrast with a background or distant source that impacts all JBB sampling locations in the same way. The mortar pit location exhibited the lowest concentrations of many pollutants; based on meteorology data, it was largely upwind of the burn pit and other local base sources. Thus, that location was used as the background location (least impacted by local sources). Concentration differences between the other two locations and the background location reflect the minimum impact of local sources on the guard tower/transportation field and housing/CASF locations (it is the minimum impact since any effect of such local sources on the background location is missed in those differences).

POLLUTANTS IN AMBIENT AIR AT JOINT BASE BALAD

The committee evaluated the CHPPM data for JBB by comparing pollutant concentrations across the site and to other locations identified in the literature, such as Beijing, China. The following section summarizes the approach and major findings of the exposure measurements taken at JBB and Appendix B discusses some of the limitations of these data. Further details on the measurements are given in Taylor et al. (2008), CHPPM and AFIOH (2009), and USAPHC (2010).

---

[2] All runway-related material was omitted, and the sky condition code CLR was replaced throughout by SKC.
[3] These data are measurements of temperature and pressure as a function of altitude, obtained using radiosonde balloons. The standard procedure is to release two balloons per day at 00:00 and 12:00 UTC (Coordinated Universal Time).

Polycyclic Aromatic Hydrocarbons

Figure 4-2 shows the average PAH concentrations observed by location and sampling campaign, using all the samples available at each location (summaries of numbers and location and timing of sampling are given in Appendix B). The overall pattern of relative concentrations remains similar for each of the sampling campaigns, although distorted in Figure 4-2 for Spring 2007 by sampling on different days at the different locations. Comparing concentration averages computed over samples taken on the same days in 2007, PAHs showed a consistent pattern of higher concentrations at the guard tower/transportation field (17/17 PAHs), lower concentrations at the H-6 housing/CASF location (11/17 PAHs, ranging from 0.82 to 1.98 times the mortar pit), and lowest concentrations at the mortar pit. This pattern suggests that the guard tower/ transportation location and the H-6 housing/CASF were affected by local sources.

The concentrations in 2009 were slightly lower than 2007, but there were differences between individual PAHs. Total PAHs were unchanged for mortar pit and H-6 housing/CASF, and they were about 30% less than 2007 values for the guard tower/transportation field.

Table 4-1 compares concentrations of PAHs at JBB with several urban locations and near or downwind of an open burning site. The concentrations measured in polluted urban areas are generally higher than the concentrations observed at JBB for most PAHs.

Particulate Matter and Metals

The JBB site, like much of the Middle East, is characterized by high concentrations of PM_{10} that are associated with windblown dust and other local and regional sources. The EPMSP found that the CHPPM 1-year Military Exposure Guideline values of 50 µg/m^3 for PM_{10} and 15 µg/m^3 for $PM_{2.5}$ were exceeded at all 15 air sampling sites in the Middle East (including JBB) for the entire 1-year air sampling period (Englebrecht et al. 2008).

The air samples taken by CHPPM at JBB in 2007 and 2009 were analyzed for PM_{10} total mass and 10 metals (antimony, arsenic, beryllium, cadmium, chromium, lead, manganese, nickel, vanadium, and zinc) within the PM_{10} sample. The average of the 90 PM_{10} measurements at JBB in 2007 was 126 µg/m^3 (range 2–535 µg/m^3); the 24-hour U.S. National Ambient Air Quality Standard (NAAQS)[4] of 150 µg/m^3 for PM_{10} was exceeded 26 out of 90 times at the three measurement locations. In 2009, the average of the 51 PM_{10} measurements (excluding those that the committee considered invalid, see Appendix B) was 709 µg/m^3 (range 104–9,576 µg/m^3) and the NAAQS was exceeded for 49 of the 51 samples. The three highest measured PM_{10} values (9,576, 2,481 and 1,951 µg/m^3) occurred on the same day during a sandstorm (USAPHC 2010). The corresponding simultaneously measured $PM_{2.5}$ concentrations were 2,662 µg/m^3, not available, and 2,889 µg/m^3 (but see Appendix B regarding measurement artifacts). There was no statistically significant difference in the average concentrations of PM_{10} or $PM_{2.5}$ among sample locations at JBB, most likely because regional contributions of windblown dust contribute the majority of the material. As discussed in Appendix B, the metal measurements were mostly "none detected," and they were not considered further by the committee.

The composition of PM was not measured in the CHPPM studies at JBB. The previous DoD EPMSP study (Engelbrecht 2008; Engelbrecht et al. 2009) attempted to measure the composition of PM at several locations in the Middle East, including at JBB and at a site near Baghdad. That study characterized PM at different sampling locations at JBB than those used by CHPPM (NRC 2010). The composition reflected a unique mixture of air pollutants and consisted of substantial amounts of windblown dust combined with elemental carbon and metals that arise from transportation and industrial activities. However, a major problem was identified in the measurements of organic carbon, rendering those results unreliable. Further, an NRC review of the EPMSP cautioned that the measurement methods used for total PM mass were subject to artifacts, that the elemental carbon results might also be affected by the problem that invalidated the organic carbon measurements, and that the x-ray fluorescence measurements of individual elements were insufficiently described to give confidence in the accuracy of the results

---

[4] 40 CFR Part 50. See http://www.epa.gov/air/criteria.html. To meet the standard, the 24-h PM_{10} should not exceed 150 µg/m^3 more than once per year on average in a 3-year period. Moreover, as pointed out in NRC report (2010), the sampling methodology used at JBB does not correspond to that required for evaluation of NAAQS compliance.

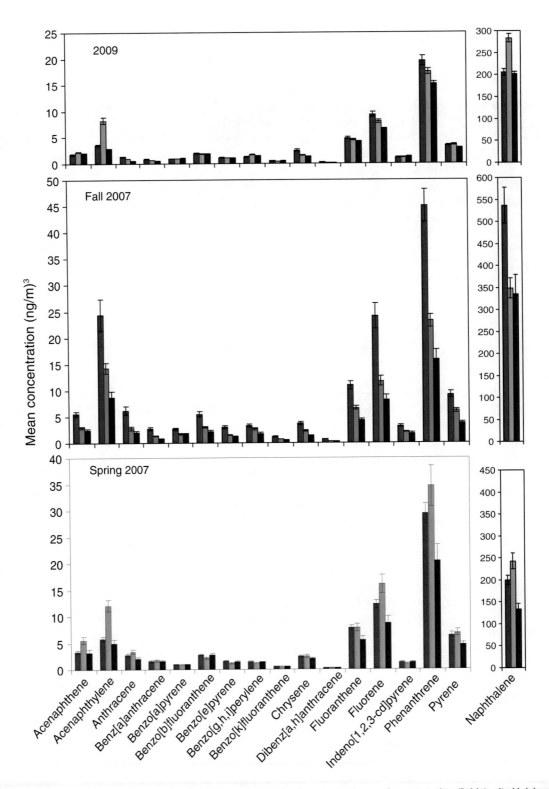

FIGURE 4-2 Mean PAH concentrations at the three sampling locations: guard tower/transportation field (red), H-6 housing/CASF (green), and mortar pit (blue) at JBB. Error bars are standard error of the mean (SEM). The y-axis scale is identical on all three panels, despite the different maximums. Naphthalene is shown separately to avoid scaling problems.

TABLE 4-1 Average Measured PAH Concentrations (ng/m^3) at JBB Compared with Measurements at Other Locations

Analyte	JBB average[a]	Rome airport apron[b]	Open burning of joss paper[c] Onsite	Open burning of joss paper[c] Down wind	Araraquara City during sugar cane burning[d]	Hong Kong[e] Urban	Hong Kong[e] Industrial
					Particulate phase only (PM$_{10}$)		
Acenaphthene	3.2	3102.4	140.3	41.3	0.7	n.d.	n.d.
Acenaphthylene	9.4	9.7	512.0	104.3	1.0	n.d.	n.d.
Anthracene	2.4	1.5	89.4	41.7	0.3	0.1	0.2
Benz[a]anthracene	1.3	1.5	39.8	30.6	n.d.	1.1	0.6
Benzo[a]pyrene	1.3	0.4	21.7	36.9	1.9	2.1	1.3
Benzo[b]fluoranthene	2.7	—	56.4	26.7	—	11.0	5.2
Benzo[e]pyrene	1.5	0.5	148.4	74.7	2.7	—	—
Benzo[g,h,i]perylene	1.8	1.9	—	—	8.5	2.9	1.2
Benzo[k]fluoranthene	0.6	—	18.8	14.5	—	0.4	0.3
Chrysene	2.2	1.1	55.8	20.7	6.6	3.4	2.0
Dibenz[a,h]anthracene	0.3	0.0	—	—	n.d.	8.0	3.2
Fluoranthene	6.3	2.9	26.0	15.9	3.3	4.8	3.6
Fluorene	11.7	10.4	56.9	61.5	0.4	0.2	0.2
Indeno[1,2,3-cd]pyrene	1.6	1.7	—	—	8.7	4.9	1.9
Naphthalene	276.5	1812.0	915.9	223.3	0.4	n.d.	n.d.
Phenanthrene	24.5	10.4	95.1	29.4	2.9	0.6	0.5
Pyrene	5.2	2.4	27.8	13.3	2.5	5.1	3.9
Number of Samples	107	5	5	5	10	11	31

NOTE: n.d. = no data.
[a]Unweighted average of locations and sampling periods from Figure 4-2 (see Table B-1, Appendix B). The JBB average and the other measurements summarized here are not necessarily representative of long-term averages.
[b]Cavallo et al. (2006).
[c]Rau et al. (2008).
[d]Godoi et al. (2004).
[e]Guo et al. (2003).

(NRC 2010). Since the same artifacts affect the PM$_{10}$ and PM$_{2.5}$ measurements taken at JBB, and since the measurements of metals were too insensitive (Appendix B) to provide useful results, the committee did not attempt to further analyze the PM and metals measurements.

Despite the potential artifacts in measurement, the PM levels measured at JBB are high compared with those found in the United States and in most urban and remote areas. For example, the EPA reports that in the United States during 2009 (the latest year for which data are available), 24-hour mean PM$_{2.5}$ concentrations were 9.9 µg/m^3 and mean PM$_{10}$ concentrations were 50.3 µg/m^3, based on 724 and 310 nationwide monitoring sites, respectively (available at http://www.epa.gov/airtrends/pm.html; accessed March 12, 2011). In Tehran, Iran, warm season mass concentrations for PM$_{10}$ were 97.6 µg/m^3 (range 76.7–122.3 µg/m^3), and 25.3 µg/m^3 for PM$_{2.5}$ (range 17.7–34.1 µg/m^3) (Halek et al. 2010). In Kuwait, summer daily average concentrations for PM$_{10}$ were 136.4 µg/m^3 (range 41.2–436.2 µg/m^3) and 55.6 µg/m^3 for PM$_{2.5}$ (range 17.6–304.4 µg/m^3) (Brown et al. 2008). Summer daily average concentrations for 2009 in Chennai, India were 76.0 ± 43.2 µg/m^3 for PM$_{10}$ and 42.2 ± 19.8 µg/m^3 for PM$_{2.5}$ (Srimuruganandam and Nagendra 2011). Additionally, in assessing air quality in the Middle East, Engelbrecht et al. (2009) found elevated levels of PM$_{2.5}$ at sites where U.S. military personnel are deployed relative to five urban areas in the United States with similar climate conditions: Las Vegas, Los Angeles, Tucson, Albuquerque, and El Paso.

The conclusions suggest that the pollutants of greatest concern at JBB may be the mixture of regional background and local sources—other than the burn pit—that contribute to high PM.

Volatile Organic Compounds

Figure 4-3 presents average concentrations of the 12 most frequently detected VOCs by location and sampling campaign, with nondetects assumed to contribute one-half the detection limit (for these VOCs, setting nondetects to zero alters the average estimates by factors between 1 and 2.4). As for PM_{10}, VOC concentrations were similar for many analytes at all the measurement locations at JBB, and there did not appear to be any consistent gradients in concentration, although differing gradients exist for some analytes at some times. This suggests that the regional background is the most important source of VOCs with intermittent local sources providing varying gradients. Table 4-2 compares the concentrations of VOCs at JBB to urban areas and reported in the literature, although some of these measurements are not directly comparable owing to different sampling methods, sampling periods, and limitations in times of day, week, or year sampled. VOC concentrations at JBB are, in general, substantially lower than polluted urban areas outside the United States.

Polychlorinated Dibenzo-Para-Dioxins/Furans

Mean concentrations for PCDD/Fs (Figure 4-4) were calculated for each of the three locations by sampling campaign. Mean concentrations of PCDDs vary considerably by site, and the differences are more pronounced than for PAHs. This spatial heterogeneity is indicative of the presence of local sources affecting these sites. PCDD/F concentrations were highest at the guard tower/transportation field location, the closest sampling location to the burn pit and downwind from it, suggesting the burn pit as a major source of PCDDs/Fs. The H-6 housing/CASF location was less affected, with PCDD/F concentrations three to four times lower than those observed at the guard tower/transportation location. The mortar pit, where individual congener concentrations were 5 to 13 times lower than those found for the guard tower/transportation field, was least affected by the PCDD/F source(s).

Table 4-3 compares the concentrations at the three JBB locations with those for an urban site in Beijing, China. On average, the concentrations at the guard tower/transportation field and H-6 housing/CASF sites were considerably higher than those reported for Beijing. In contrast, the PCDD/F concentrations observed at the mortar pit background location were similar to or lower than those reported for Beijing. The comparisons in Table 4-3 include only one external study because few published studies report all the 2,3,7,8-substituted PCDD/F congeners. The majority of such studies report data in 2,3,7,8-TCDD toxic equivalent concentration units (I-TEQ 1989—International Toxicity Equivalents, 1989 method, with Toxic Equivalency Factor [TEF] given in Table 4-4).[5]

Table 4-5 shows TEQ values averaged over the three sampling campaigns (Spring 2007, Fall 2007, and 2009) for each of the three JBB sites, and compares them with values from a variety of environments around the world. TEQ values determined for the guard tower/transportation field and H-6 housing/CASF sites are considerably higher than those reported in other environmental studies, except for a landfill fire in Zagreb, Croatia. Finally, the average TEQ value estimated for the mortar pit site was seven times lower than the one determined for the guard tower/field and somewhat lower than that obtained for the Beijing urban site, but higher than values reported for most rural and suburban environments around the world.

There are significant differences in total PCDD/F concentration measurements between the three sampling locations in 2007 (see Table 4-5); the guard tower/transportation field total PCDD/F concentration is 7.5 times the mortar pit concentration, and the H-6 housing/CASF total PCDD/F concentration is 2.5 times the mortar pit concentration. In the 2009 measurements, the total PCDD/F concentrations are decreased by differing factors from 2007; the guard tower/transportation field total PCDD/F concentration is only 30% of that in 2007, while the H-6 housing/CASF and mortar pit total PCDD/F concentrations are 65% of their values in 2007. Also, the guard tower/transportation field total PCDD/F concentration is about 5 times the mortar pit concentration and the H-6 housing/CASF total PCDD/F concentration is still about 2.5 times the mortar pit concentration.

Table 4-5 shows that the measured average concentrations fall with the estimated burn rate, although they are not proportional to the burn rates. However, such lack of proportionality is not surprising in view of the large

---

[5] The I-TEF values have since been updated, but the 1989 version is used here to allow comparison with the literature. Generally, congener distributions are not published so the committee could not recalculate to more recent TEF values.

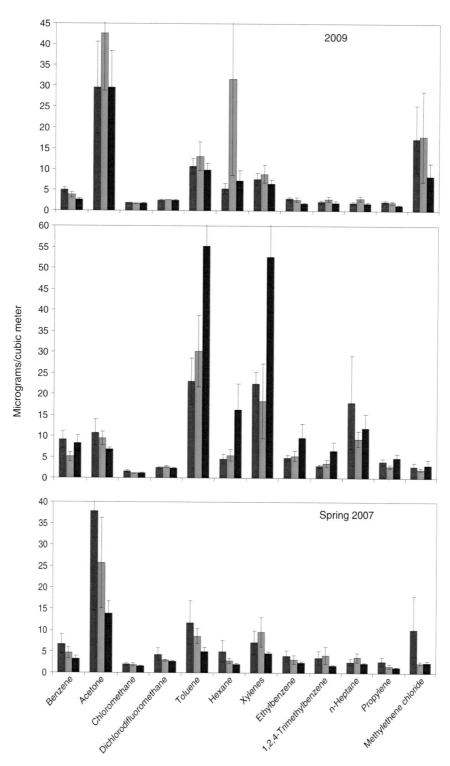

FIGURE 4-3 Mean VOC concentrations at the three sampling locations: guard tower/transportation field (red), H-6 housing/CASF (green), and mortar pit (blue) at Joint Base Balad.
NOTE: Nondetects are treated as one-half the detection limit; error bars are SEM. The y-axis scales are identical on the three panels, despite their different maximums.

TABLE 4-2 Comparison of Average Measured Concentrations ($\mu g/m^3$) of the 12 Most Frequently Detected VOCs at JBB with Other Locations

Analyte	JBB[a]	Taiwan[b]	Bangkok, industrial-commercial[c]	Bangkok, commercial-residential[c]	Karachi, urban[d]	Karachi, roadside[d]	Athens center[e]	Los Angeles, CA[f]
Benzene	5.4	4.2	56.0	25.0	16.7	63.1	16.0	1.6
Acetone	22.9	nm	nm	nm	nm	nm	nm	21.1
Chloromethane	1.6	nm	nm	nm	5.6	4.0	nm	nm
Dichlorodifluoromethane	2.8	nm	nm	nm	3.2	4.7	nm	nm
Toluene	18.6	27.6	279.0	72.0	26.8	166.3	54.1	5.4
Hexane	8.9	2.5	68.0	40.0	26.5	251.0	5.7	nm
Total Xylenes	15.3	11.3	150.0	73.0	18.3	200.4	68.8	4.3
Ethylbenzene	4.1	5.2	27.0	13.0	nm	nm	11.8	0.9
n-Heptane	3.2	nm	53.0	24.0	16.0	160.3	9.9	nm
1,2,4-Trimethylbenzene	6.0	12.8	28.0	14.0	4.9	32.1	19.2	nm
Propylene	2.5	9.5	nm	nm	9.5	67.3	6.7	nm
Methylene chloride	7.4	nm	nd	nd	1.1	0.6	nm	1.1
Number of samples	122	40	1	1	50	28	12	29

NOTE: nd = not detected; nm = not measured.
[a]Unweighted average across locations and times from Figure 4-3 (see Table B-3, Appendix B). The JBB average and the other measurements summarized here are not necessarily representative of long-term averages.
[b]Hsieh and Tsai (2003).
[c]Kungskulniti and Edgerton (1990).
[d]Barletta et al. (2002).
[e]Moschonas and Glavas (1996).
[f]CARB (2009).

variability observed in emissions from open burning (Lemieux et al. 2003, 2004) and the possibility that the waste streams differed between those time periods.

SUMMARY AND IMPLICATIONS FOR EXPOSURE

Ambient air concentrations of PCDD/Fs, PAHs, VOCs, and PM were measured at JBB, and the committee used these values to estimate the impact of the burn pit on air pollution at JBB. Of the three monitoring locations at JBB, the mortar pit was considered a background site and was located upwind of the burn pit, the other two locations—H-6 housing/CASF and the guard tower/transportation field—were considered to be downwind of the burn pit; the guard tower/transportation field location was closest in proximity to the burn pit. Ambient air data were evaluated for composition and concentration at each of the sites to determine differences that may be attributed to the burn pit or other known sources. Following are the conclusions of these analyses:

- Background ambient air concentrations of PM at JBB are high, with average concentrations above the U.S. air pollution standards. The high background PM concentrations are most likely derived from local sources, such as traffic and jet emissions as well as regional sources, including long-range anthropogenic emissions and dust storms, although emissions from the burn pits may contribute a small amount of PM.
- PCDDs/Fs were detected at low concentrations in nearly all samples, and the burn pit was likely the major source of these chemicals. The toxic equivalents of the concentrations are high compared with locations in the United States and even with polluted urban environments worldwide, but they are below those associated locally with individual sources.
- Ambient VOC and PAH concentrations were similar to those reported for polluted urban environments outside the United States, and the major sources of these pollutants are regional background, ground transportation, stationary power generation, and the JBB airport.

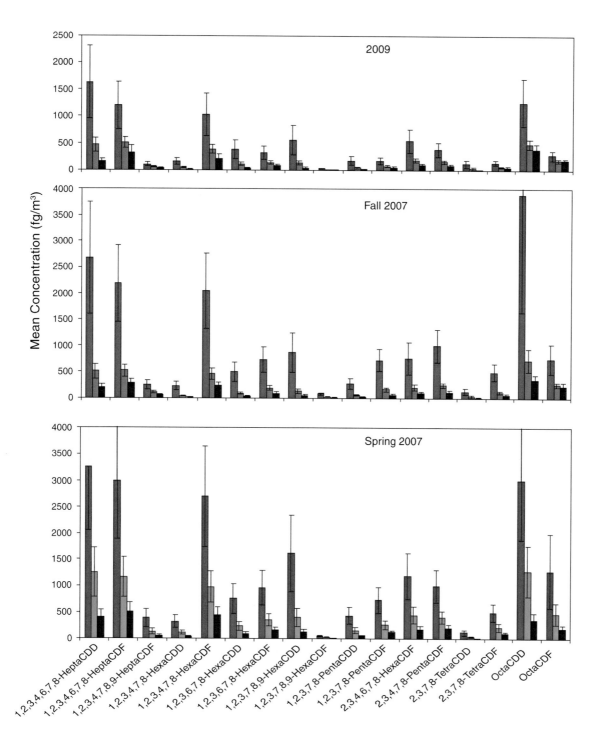

FIGURE 4-4 Mean PCDD/F concentrations (fg/m^3) at the three sampling locations by sampling campaign: guard tower/transportation field (red), H-6 housing/CASF (green), and mortar pit (blue) at JBB.
NOTE: Nondetects were treated as one-half the detection limit; error bars are SEM. The y-axis scale is identical on all three panels, despite the different maximums.

TABLE 4-3 Measured PCDD/F Concentrations (fg/m^3) at JBB Compared with Beijing, China

Congener	ITEF/89[a]	2006	Guard tower/transportation field			H-6 Housing/CASF			2007 Mortar pit		2009	Beijing[b]
			Spring 2007	Fall 2007	2009	Spring 2007	Fall 2007	2009	Spring 2007	Fall 2007		
1,2,3,4,6,7,8-HeptaCDD	0.01	1416	3266	2677	1632	1262	511	460	397	204	152	240
1,2,3,4,6,7,8-HeptaCDF	0.01	1638	2983	2180	1196	1168	523	509	511	294	320	926
1,2,3,4,7,8,9-HeptaCDF	0.01	181	387	244	103	137	107	53	58	59	37	155
1,2,3,4,7,8-HexaCDD	0.1	136	322	228	153	115	39	44	42	19	16	19
1,2,3,4,7,8-HexaCDF	0.1	1361	2697	2047	1028	982	467	386	440	232	209	241
1,2,3,6,7,8-HexaCDD	0.1	290	755	500	380	241	88	109	92	38	29	38
1,2,3,6,7,8-HexaCDF	0.1	487	962	739	323	362	192	137	170	95	78	237
1,2,3,7,8,9-HexaCDD	0.1	428	1623	870	553	404	133	127	124	49	37	28
1,2,3,7,8,9-HexaCDF	0.1	37	54	80	28	28	23	14	11	14	12	90
1,2,3,7,8-PentaCDD	0.5	192	434	282	170	153	56	41	50	22	18	23
1,2,3,7,8-PentaCDF	0.05	365	730	725	168	263	165	71	119	60	47	151
2,3,4,6,7,8-HexaCDF	0.1	594	1187	762	539	449	199	181	184	96	85	305
2,3,4,7,8-PentaCDF	0.5	629	1000	1001	378	406	250	153	202	112	78	249
2,3,7,8-TetraCDD	1	58	118	126	117	39	35	37	11	15	8	7
2,3,7,8-TetraCDF	0.1	260	493	497	135	217	107	59	103	58	45	144
OctaCDD	0.001	1338	3006	3887	1258	1279	719	486	360	354	384	602
OctaCDF	0.001	636	1286	751	271	471	255	181	187	229	173	902
TEQ/89		880	1752	1431	745	639	333	254	270	151	115	275

NOTE: ITEF = International toxicity equivalency factor; TEQ = toxic equivalent.
[a]EPA (1989).
[b]Li et al. (2008).

TABLE 4-4 PCDD/F Concentrations in TEQ Units at JBB in 2007 Compared with Other Locations

Location	fg I-TEQ/m^3 (I-TEQ/89 method)	Reference
Balad, Guardtower/Transportation Field	1,309	
Balad, H-6 Housing/CASF	409	
Balad, Mortar Pit	179	
Catalonia, Spain, industrial, 1994–2004	140	1
Catalonia, Spain, rural, 1994–2004	28	1
Athens, Greece, background, July 2000	8	1
Catalonia, Spain, traffic, 1994–2004	72	1
Beijing, China, 3 districts, Feb–Dec 2006	275	1
Athens, Greece, urban, July 2000	42	1
Porto, Portugal, Suburban, 1999–2004	149	1
Lisbon, Portugal, Suburban, 1999–2004	34	1
Madeira, Portugal, rural, 1999–2004	15	1
Zagreb, Croatia, May 1997–March 2000	61	2
Zagreb, Croatia, during garden waste fire	90	2
Zagreb, Croatia, during landfill fire	13,200	2

NOTE: The JBB averages and the other measurements summarized here are not necessarily representative of long-term averages. References are (1) Li et al. (2008), and (2) Krauthacker et al. (2006).

TABLE 4-5 Average Total 2,3,7,8-PCDD/F Concentrations (Sum of All 2,3,7,8 Congeners, in fg/m^3) by Sampling Location and Period

Sampling Date	Burn Rate at Pit, tons*	Guardtower/ Transportation Field	H-6 Housing/ CASF	Mortar pit
Spring 2007	100–200	21,303	7,975	3,063
Fall 2007	50–100	17,593	3,868	1,949
2009	10	8,434	3,047	1,727

*USAPHC (2010).

These conclusions are based on the measurements available, but those measurements omitted some of the pollutants considered criteria pollutants in the United States, such as sulfur dioxide, ozone, nitrogen dioxide, and carbon monoxide. Since the burn pit is likely to have been a source of some of those pollutants, the evaluation of air monitoring data alone cannot provide a complete picture of the potential effects of burn pit emissions. Furthermore, there likely were additional pollutants emitted from the burn pit that were not measured during the CHPPM monitoring campaigns, since the burning of household waste is known to emit other pollutants (EPA 1997, 2001; Lemieux et al. 2003, 2004). Various modeling efforts were undertaken by the committee to examine the consistency of its conclusions, including air dispersion modeling, PMF analysis, and scale-up from emissions observed in experimental burning of household waste in barrels. These modeling efforts were largely consistent with the committee's conclusions, although they were limited by the available data. However, it is unlikely that the measurements presented here misrepresent the general trend of low contributions of emissions from the burn pit at the monitoring sites at JBB. The potential health risks associated with these exposures are discussed further in Chapter 5.

REFERENCES

Barletta, B., S. Meinardi, I. J. Simpson, H. A. Khwaja, D. R. Blake, and F. S. Rowland. 2002. Mixing ratios of volatile organic compounds (VOCs) in the atmosphere of Karachi, Pakistan. *Atmospheric Environment* 36(21):3429-3443.

Brown, K. W., W. Bouhamra, D. P. Lamoureux, J. S. Evans, and P. Koutrakis. 2008. Characterization of particulate matter for three sites in Kuwait. *Journal of the Air & Waste Management Association* 58(8):994-1003.

CARB (California Air Resources Board). 2009. *Annual toxics summary by monitoring site for Los Angeles-North Main Street 2009.* http://www.arb.ca.gov/adam/toxics/sitesubstance.html (accessed August 16, 2011).

Cavallo, D., C. L. Ursini, G. Carelli, I. Iavicoli, A. Ciervo, B. Perniconi, B. Rondinone, M. Gismondi, and S. Iavicoli. 2006. Occupational exposure in airport personnel: Characterization and evaluation of genotoxic and oxidative effects. *Toxicology* 223(1-2):26-35.

CHPPM (U.S. Army Center for Health Promotion and Preventive Medicine) and AFIOH (U.S. Air Force Institute for Operational Health). 2009. *Addendum 2. Screening health risk assessment burn pit exposures Balad Air Base, Iraq, May 2008.* USACHPPM Report No. 47-MA-08PV-08/AFIOH Report No. IOH-RS-BR-TR-2008-0001. Aberdeen Proving Ground, MD: U.S. Army Center for Health Promotion and Preventive Medicine. August.

Engelbrecht, J. P. 2008. *Department of Defense Enhanced Particulate Matter Surveillance Program.* Reno, NV: Desert Research Institute.

Engelbrecht, J. P., E. V. McDonald, J. A. Gillies, R. K. M. Jayanty, G. Casuccio, and A. W. Gertler. 2009. Characterizing mineral dusts and other aerosols from the Middle East—Part 1: Ambient sampling. *Inhalation Toxicology* 21(4):297-326.

EPA (U.S. Environmental Protection Agency). 1989. *Interim procedures for estimating risks associated with exposures to mixtures of chlorinated dibenzo-p-dioxins and –dibenzofurans (CDDs and CDFs) and 1989 update.* EPA/625/3-89/016. Washington, DC: U.S. Environmental Protection Agency. March.

EPA. 1997. *Evaluation of emissions from the open burning of household waste in barrels.* Vol 1. EPA/600/R-97-134a. Research Triangle Park, NC: U.S. Environmental Protection Agency.

EPA. 2001. *Dioxin emission database.* EPA/600/C-01/012. http://cfpub.epa.gov/ncea/cfm/recordisplay.cfm?deid=20797 (accessed September 23, 2010).

Godoi, A. F., K. Ravindra, R. H. Godoi, S. J. Andrade, M. Santiago-Silva, L. Van Vaeck, and R. Van Grieken. 2004. Fast chromatographic determination of polycyclic aromatic hydrocarbons in aerosol samples from sugar cane burning. *Journal of Chromatography A* 1027(1-2):49-53.

Guo, H., S. C. Lee, K. F. Ho, X. M. Wang, and S. C. Zou. 2003. Particle-associated polycyclic aromatic hydrocarbons in urban air of Hong Kong. *Atmospheric Environment* 37(38):5307-5317.

Gullett, B. K., B. Wyrzykowska, E. Grandesso, A. Touati, D. G. Tabor, and G. S. Ochoa. 2010. PCDD/F, PBDD/F, and PBDE emissions from open burning of a residential waste dump, Table S-1. *Environmental Science and Technology* 44(1):394-399.

Halek, F., M. Kianpour-Rad, and A. Kavousirahim. 2010. Seasonal variation in ambient PM mass and number concentrations (case study: Tehran, Iran). *Environmental Monitoring and Assessment* 169(1-4):501-507.

Hsieh, C. C., and J. H. Tsai. 2003. VOC concentration characteristics in Southern Taiwan. *Chemosphere* 50(4):545-556.

Krauthacker, B., S. H. Romanic, M. Wilken, and Z. Milanovic. 2006. PCDD/Fs in ambient air collected in Zagreb, Croatia. *Chemosphere* 62(11):1829-1837.

Kungskulniti, N., and S. A. Edgerton. 1990. Ambient volatile organic-compounds at selected sites in Bangkok-City, Thailand. *Chemosphere* 20(6):673-679.

Lemieux, P. M., B. K. Gullett, C. C. Lutes, C. K. Winterrowd, and D. L. Winters. 2003. Variables affecting emissions of PCDD/Fs from uncontrolled combustion of household waste in barrels. *Journal of the Air & Waste Management Association* 53(5):523-531.

Lemieux, P. M., C. C. Lutes, and D. A. Santoianni. 2004. Emissions of organic air toxics from open burning: A comprehensive review. *Progress in Energy and Combustion Science* 30(1):1-32.

Li, Y. M., G. B. Jiang, Y. W. Wang, Z. W. Cai, and Q. H. Zhang. 2008. Concentrations, profiles and gas-particle partitioning of polychlorinated dibenzo-p-dioxins and dibenzofurans in the ambient air of Beijing, China. *Atmospheric Environment* 42(9):2037-2047.

Moschonas, N., and S. Glavas. 1996. C_3-C_{10} hydrocarbons in the atmosphere of Athens, Greece. *Atmospheric Environment* 30(15):2769-2772.

NRC (National Research Council). 2003. *Toxicologic assessment of jet-propulsion fuel 8.* Washington, DC: The National Academies Press.

NRC. 2010. *Review of the Department of Defense Enhanced Particulate Matter Surveillance Program report.* Washington, DC: The National Academies Press.

Srimuruganandam, B., and S. M. S. Nagendra. 2011. Characteristics of particulate matter and heterogeneous traffic in the urban area of India. *Atmospheric Environment* 45(18):3091-3102.

Taylor, G., V. Rush, A. Deck, and J.A. Vietas. 2008. *Screening health risk assessment burn pit exposures, Balad Air Base, Iraq and Addendum Report*. IOH-RS-BR-TR-2008-0001/U.S.ACHPPM 47-MA-08PV-08. Brooks City-Base, TX: Air Force Institute for Operational Health and U.S. Army Center for Health Promotion and Preventative Medicine. May.

USAF (U.S. Air Force). 2009. Surface weather observations. *Air Force Manual 15-111*. Arlington, VA: U.S. Air Force. March 10.

USAPHC (U.S. Army Public Health Command). 2010. *Screening health risk assessments, Joint Base Balad, Iraq, 11 May–19 June 2009*. Aberdeen Proving Ground, MD: U.S. Army Center for Health Promotion and Preventive Medicine. July.

5

Health Effects of Air Pollutants Detected at Joint Base Balad

As discussed in Chapter 4, the committee found the air monitoring data gathered at Joint Base Balad (JBB) in 2007 and 2009 to be useful for identifying the major air pollutants and their sources in and around JBB. In Chapter 5, the committee begins its assessment of the potential long-term health effects associated with exposure to those air pollutants. To this end, the committee (1) reviews the health effects associated with the air pollutants identified in Chapter 4; (2) lists the health effects associated with exposure to the most frequently detected pollutants at JBB regardless of their source; and (3) provides a qualitative analysis of the assembled health effects data from a chemical mixture or cumulative risk perspective. The committee did not conduct a quantitative risk assessment. Exposure to both burn pit emissions and air pollutants from other sources in and around JBB will likely be of concern in future epidemiologic studies.

RATIONALE AND DATA SOURCES

The committee considers several of the air pollutants highlighted in Chapter 4 to be of concern because of their association with burn pit emissions (dioxins and dioxin-like compounds) and because some of the concentrations exceeded U.S. air quality standards (for example, particulate matter [PM]) or were in excess of concentrations found in polluted urban environments worldwide. Health effects associated with these pollutants are well documented. The committee drew on previous expert panel reviews and selected research literature to summarize the health effects of the chemicals of concern.

Typically, the hazard identification step of a risk assessment addresses what the chemicals of concern are as well as the specific health effects associated with exposure to them (see Figure 3-1) (NRC 1983, 2009). The screening health risk assessments conducted by the U.S. Army Center for Health Promotion and Preventive Medicine (CHPPM, now the U.S. Army Public Health Command) and the U.S. Air Force Institute for Operational Health (AFIOH) focused on those chemicals detected in the air monitoring campaigns but restricted their review of associated health effects to the broad categories of cancer and either to noncancer effects in general (Taylor et al. 2008), or to just the primary target organs for noncancer effects (CHPPM 2009, USAPHC 2010), and therefore specific health effects potentially related to exposure to air pollutants at JBB were not fully presented. The committee assembled specific health effects data (including all target organs) on the detected air pollutants as a step towards identifying the potential long-term effects of them.

One step in the committee's analysis of the air monitoring data was to evaluate how often a particular pollutant was detected among the samples taken (see Chapter 4, Table 4-6). The pollutants listed in Table 5-1 were detected in at least 5% of the air monitoring samples collected at JBB in 2007 and 2009 (n = 47 chemicals). There are an additional four pollutants (1,2,4-trichlorobenzene, 1,3-dichlorobenzene, 1,3-butadiene, and 1,2-dichlorobenzene) that were detected at JBB although in fewer than 5% of the samples, but they were included in the committee's assessment because they are expected to be present in burn pit emissions on the basis of burn barrel experiments (Lemieux et al. 2003, 2004; see also Chapter 4, Table 4-6). Health effects of particulate matter, dioxins (as represented by 2,3,7,8-tetrachlorodibenzo-*p*-dioxin [TCDD]), and metals detected at JBB (lead, zinc, and antimony) are also described. In all, 56 pollutants are profiled in Table 5-1.

When available, specific cancer and noncancer health effects data for the pollutants in Table 5-1 were obtained from the U.S. Environmental Protection Agency's (EPA's) Integrated Risk Information System (IRIS), Agency for Toxic Substances and Disease Registry (ATSDR) Toxicological Profiles, the National Institute of Occupational Safety and Health (NIOSH), or the National Library of Medicine's (NLM's) Hazardous Substance Data Bank. IRIS is the source of much of the toxicity information discussed in this chapter. The toxicity values and supporting documentation developed by the EPA and other agencies are the result of extensive review and synthesis of health effects literature and are designed for practical application in assessments of human health risks. The committee recognizes there are concerns regarding IRIS (NRC 2009); nevertheless, IRIS and other agency databases provide the best readily available evaluation of health effects from exposure to toxic substances.

HEALTH EFFECTS OF SELECTED AIR POLLUTANTS DETECTED AT JBB

The evaluation of air monitoring data from JBB reported in Chapter 4 indicates that combustion products from burn pits were associated with low concentrations of dioxins and dioxin-like compounds but contributed a relatively small proportion of PM compared to local dust and other sources. The committee recognized that personnel at military bases similar to JBB were exposed to many hazardous agents associated with adverse health effects in addition to burn pits. These exposures may result from use of kerosene heaters, JP-8 fuel, and tobacco products in addition to the hazards and stress inflicted by war. Assessment of these additional exposures was outside the committee's scope and thus focus was on only exposures related to burn pits. While not measured directly, first hand descriptions of the burn pits describe volumes of smoke resulting from the burn pit and use of JP-8 fuel to encourage combustion (see Chapter 2). Both smoke and JP-8 fuel are associated with adverse health effects as described below.

JP-8 and similar fuels were used by the military to power aircraft, ground vehicles, tent heaters, and cooking stoves. These fuels were also used for less conventional purposes, such as suppressing sand, cleaning equipment, and burning trash. Military personnel serving in the Gulf War theater of operations could have been exposed to the uncombusted fuels, the combustion products from the burning of those fuels, or a combination of uncombusted and combusted materials (IOM 2005). Health effects of JP-8 are similar to those of kerosene, the primary component of JP-8, exposures generally causing nervous system effects. Large doses of inhaled JP-8 are known to cause headaches and fatigue, and affect concentration and coordination, while more chronic exposures can affect sleep, motivation, and cause dizziness, but not cancer (ATSDR 1998). As part of the IOM's continuing series of *Gulf War and Health* reports, a previous IOM committee assessed the toxicological and epidemiological effects of these fuels and their combustion products. That committee did not find any association between exposure to uncombusted fuels and long-term health effects. Conversely, fuel combustion products were found to have sufficient evidence of an association with lung cancer; limited suggestive evidence of an association with several other cancers (nasal cavity, nasopharynx, oral cavity, laryngeal, and bladder cancers), reproductive effects, and incident asthma (IOM 2005). (See Chapter 6 for a description of the categories of association used for these combustion products and health effects.)

While products of combustion vary greatly based on fuel composition and conditions of the burn, several health effects have been described consistently in association with exposure to smoke. Studies have examined health effects from exposure to ambient air pollution, exposure to wood smoke from indoor wood-burning stoves and fireplaces, and exposure to smoke from wildland or agricultural fires. Wood smoke has been associated with premature death, chronic obstructive pulmonary disease (COPD), tuberculosis, acute lower respiratory infections, asthma and respiratory symptoms, asthma related hospital admissions and emergency room visits, decreased lung function, and in some studies cardiac hospital admissions (Boman et al. 2003, 2006; Naeher et al. 2007). Asthma symptoms, asthma related hospital admissions, and cough are shown to be related to PM_{10} in five studies that specified wood smoke as a major contributor to ambient air pollution (Boman et al 2003).

Naeher et al. (2007) make the point that in addition to wood and biomass, tobacco, the most well-studied biomass smoke, is also important in determining and apportioning health effects from overall smoke exposure. Tobacco smoke provides another example of demonstrated adverse health effects from smoke and is especially relevant to military personnel because prevalence of smoking is elevated in military populations (IOM 2009). Tobacco smoke contains many environmental contaminants, including particulate matter, acrolein, polycyclic aromatic hydrocarbons, benzene, and metals. The 2004 Surgeon General's report associated tobacco smoke with cancer, particularly of the lung and larynx, as well as the urinary tract and oral cavity; cardiovascular disease, including acute myocardial infarction, angina, stroke, and peripheral artery disease; pulmonary disease such as chronic bronchitis, emphysema, asthma, and increased susceptibility to pneumonia and other respiratory infections; gastrointestinal disease such as peptic ulcer and esophageal reflux; and reproductive effects, including low birth weight, spontaneous abortion, premature birth, and reduced fertility (U.S. Surgeon General 2004). Even exposure to secondhand smoke can result in long-term health effects, in particular, an increased risk for lung cancer (IARC 2004) and cardiovascular disease, including death and acute myocardial infarction (IOM 2010).

The committee acknowledges the many occupational and environmental exposures present at JBB; however, direct information on other exposures is lacking and outside the task of this report. Thus, the committee focused on the specific pollutants determined to be associated with burn pits—polycyclic aromatic hydrocarbons (PAHs), volatile organic compounds (VOCs), dioxins/furans, and PM—even though the proportion contributed by burn pits is thought to be relatively small, with the exception of dioxin.

Dioxins and Dioxin-Like Compounds

The dioxin TCDD is classified as carcinogenic to humans by the EPA (2003a) and by the International Agency for Research on Cancer (1997). In the 2003 draft report *Exposure and Human Health Reassessment of 2,3,7,8-Tetrachlorodibenzo-p-Dioxin and Related Compounds*, the EPA focused on three epidemiologic cohort studies—Ott and Zober (1996); Becher et al. (1998); and Steenland et al. (2001)—that provided quantitative dose-response estimates that linked serum dioxin levels to cancer mortality (NRC 2006). The IARC also evaluated the study by Ott and Zober (1996) and additional studies by Fingerhut et al. (1991), Becher et al. (1996), Hooiveld et al. (1996), and Steenland et al. (2004). The cohort studies reviewed in these evaluations were used principally because they included subjects with serum dioxin levels higher than background and who were in industrial settings, which allowed for better characterization of exposure. The classification of dioxin as carcinogenic to humans addresses total cancer mortality and does not specify tumor type. The focus on total cancers is a result of the presumption that TCDD is not in itself genotoxic, but that rather it acts primarily as a promoter rather than an initiator of cancer (IARC 1997). The *Veterans and Agent Orange: Update 2008* (VAO) from the Institute of Medicine (IOM) continued to support the association between exposure of Vietnam veterans to the TCDD-contaminated Agent Orange and soft-tissue sarcoma, non-Hodgkin's lymphoma, Hodgkin's disease, and chronic lymphocytic leukemia. The IOM report also broadened the categorization of sufficient evidence of an association between Agent Orange exposure and health effects to cover chronic lymphocytic leukemia, including hairy-cell leukemia and other chronic B-cell leukemias (IOM 2008).

In animal studies with oral administration, TCDD exposure has been associated with noncancer health effects. High exposures to TCDD affect many organs and can result in organ dysfunction and death. Other reported specific adverse health effects include diabetes; immunologic response; altered neurologic function; reproductive and developmental effects including birth defects; changes to the endocrine system; and wasting syndrome, which results in the loss of adipose and muscle tissues and severe weight loss (Mandal 2005; IOM 2008; White and Birnbaum 2009).

Rats and mice exposed to TCDD had increased incidence of degenerative cardiovascular lesions, cardiomyopathy, chronic active arteritis, increased heart weight, increased blood pressure, and severe atherosclerotic lesions (Humblet et al. 2008). The 2008 VAO update committee, after extensive deliberation regarding the strengths and

TABLE 5-1 Long-Term Health Effects for Chemicals of Interest Detected at JBB

Chemical Name and CAS Number	Class	Long-Term Health Effects[a]	Inhalation Unit Risk (mg/m^3)	Reference Concentration (mg/m^3)[c]	1-yr Air MEG (mg/m^3)[d]
Acenaphthene 83-32-9	PAH	Increased liver weight, increased cholesterol, vascular disorders, and degeneration in the internal organs and central nervous system	NA	NA	1.40E-01
Acenaphthylene 208-96-8	PAH	NA	NA	NA	2.80E-02
Anthracene 120-12-7	PAH	No observed effects at highest dose	NA	NA	3.50E+01
Benz[a]anthracene 56-55-3	PAH	Probable carcinogen,[b] lung and liver cancers	1.1E-07 C	NA	5.40E-02
Benzo[a]pyrene 50-32-8	PAH	Probable carcinogen,[b] stomach and respiratory tract tumors	1.3E-06 C	NA	5.40E-03
Benzo[b]fluoranthene 205-99-2	PAH	Probable carcinogen,[b] lung and skin tumors, effects on liver	1.1E-07 C	NA	5.40E-02
Benzo[e]pyrene 192-97-2	PAH	NA	NA	NA	NA
Benzo[g,h,i]perylene 191-24-2	PAH	NA	NA	NA	NA
Benzo[k]fluoranthene 207-08-9	PAH	Probable carcinogen,[b] lung and skin tumors	1.1E-07 C	NA	5.40E-01
Chrysene 218-01-9	PAH	Probable carcinogen,[b] carcinomas and malignant lymphoma	1.1E-08 C	NA	5.50E+00
Dibenz[a,h]anthracene 53-70-3	PAH	Probable carcinogen,[b] stomach and respiratory tract tumors	1.2E-06 C	NA	5.40E-03
Fluoranthene 206-44-0	PAH	Nephropathy, increased liver weights, hematological alterations	NA	NA	1.40E+00
Fluorene 86-73-7	PAH	Blood effects; increased liver, spleen, and kidney weights	NA	NA	1.40E+00
Indeno[1,2,3-cd]pyrene 193-39-5	PAH	Probable carcinogen,[b] lung and skin tumors	1.1E-07 C	NA	5.40E-02
Naphthalene 91-20-3	PAH	Possible carcinogen,[b] respiratory tumors; decreased body weights	3.4E-08 C	3.0E-03 I	7.10E-02
Phenanthrene 85-01-8	PAH	NA	NA	NA	4.20E-02
Pyrene 129-00-0	PAH	Nephropathy and decreased kidney weight	NA	NA	1.05E-01
1,2-Dichlorobenzene 95-50-1	VOC	Increased liver and kidney weight, liver necrosis, renal tubular degeneration	NA	2.0E-01 H	1.40E+00

Chemical / CAS	Category	Effects			
1,2,4-Trimethylbenzene 95-63-6	VOC	Long term exposure in workers: defats the skin, lungs may be affected, chronic bronchitis, CNS (impaired neurobehavioral test performance), hypochromic anemia (NLM 2011)	NA	7.0E-03 P	3.06E+00
1,3,5-Trimethylbenzene 108-67-8	VOC	Long term exposure in workers: defats the skin, lungs may be affected, chronic bronchitis, CNS (impaired neurobehavioral test performance), hypochromic anemia (NLM 2011)	NA	NA	3.06E+00
1,4-Dichlorobenzene 106-46-7	VOC	Increased liver and kidney weights, liver tumors	1.1E-08 C	8.0E-01 I	1.70E+00
2-Butanone (MEK) 78-93-3	VOC	Maternal and developmental toxicity (e.g., decreased weight gain in dams and decreased body weight and skeletal variations in pups)	NA	5.0E+00 I	1.44E+01
4-Ethyltoluene 622-96-8	VOC	NA	NA	NA	NA
Acetone 67-64-1	VOC	Eye and respiratory tract irritation, neurobehavioral and neurological effects (e.g., reduced nerve conduction velocity, increased reaction time)	NA	3.1E+01 A	2.90E+01
Acrolein 107-02-8	VOC	Respiratory and inflammatory responses, nasal lesions, increased heart and kidney weights, liver necrosis, decreased body weight gain	NA	2.0E-05 I	1.40E-05
Benzene 71-43-2	VOC	Known carcinogen,[b] leukemia and hematologic neoplasms; progressive deterioration of hematopoietic function with chronic exposure, suppression of circulating B-lymphocytes, menstrual disorders, limited evidence of reproductive toxicity and neurotoxicity	7.8E-09 I	3.0E-02 I	3.90E-02
1,3-Butadiene 106-99-0	VOC	Probable carcinogen,[b] liver, lung, ovary, and mammary tumors; lymphohematopoietic cancers and leukemia; reproductive and developmental effects (e.g., ovarian and testicular atrophy, fetal skeletal variations, decreased fetal weight)	3.0E-08 I	2.0E-03 I	1.70E-02
Carbon disulfide 75-15-0	VOC	Peripheral nervous system dysfunction (e.g., reduced nerve conduction velocity), possible CNS and ocular effects (e.g., blurred vision, memory difficulty)	NA	7.0E-01 I	4.80E-01
Chlorodifluoromethane 75-45-6	VOC	Increased kidney, adrenal and pituitary weights	NA	5.0E+01 I	3.42E+00

continued

TABLE 5-1 Continued

Chemical Name and CAS Number	Class	Long-Term Health Effects[a]	Inhalation Unit Risk (mg/m^3)	Reference Concentration (mg/m^3)[c]	1-yr Air MEG (mg/m^3)[d]
Chloromethane 74-87-3	VOC	Cerebellar lesions, central nervous system dysfunction	NA	9.0E-02 I	2.70E+00
Cyclohexane 110-82-7	VOC	Developmental and reproductive toxicity (reduced maternal and pup body weights), CNS depression	NA	6.0E+00 I	NA
Dichlorodifluoromethane 75-71-8	VOC	Cardiovascular system and peripheral nervous system effects (CDC 2010)	NA	2.0E-01 H	9.90E+01
Ethylbenzene 100-41-4	VOC	Increased liver, kidney and spleen weights; developmental toxicity (e.g., skeletal variations)	2.5E-06 C	1.0E+00 I	3.00E+00
Hexane 110-54-3	VOC	Peripheral neuropathy	NA	7.0E-01 I	4.30E+00
Isooctane 540-84-1	VOC	NA	NA	NA	NA
Isopropyl alcohol 67-63-0	VOC	Eye and respiratory tract irritation: increased liver enzymes and relative liver weight; narcosis at highest exposures (CDC 2010; NLM 2011)	NA	7.0E+00 C	NA
Methyl tert-butyl ether (MtBE) 1634-04-4	VOC	Increased absolute and relative liver and kidney weights and increased severity of spontaneous renal lesions (females), increased prostration (females), and swollen periocular tissue (males and females)	2.67E-10 C	3.0E+00 I	2.10E+00
Methylene chloride 75-09-2	VOC	Probable carcinogen,[b] liver, mammary gland, salivary gland, lung tumors; liver toxicity (e.g., fatty changes)	4.7E-10 I	1.0E+00 A	2.10E+00
n-Heptane 142-82-5	VOC	Skin, eye and respiratory irritant, and CNS depression at high exposures (CDC 2010; NLM 2011)	NA	NA	NA
Octane 111-65-9	VOC	Skin, eye, and respiratory irritant, and CNS depression at high exposures (CDC 2010; NLM 2011)	NA	NA	NA
Pentane 109-66-0	VOC	Skin, eye, and respiratory irritant, and CNS depression at high exposures (CDC 2010; NLM 2011)	NA	1.0E+00 P	NA

Chemical	Type	Health Effects	Col 4	Col 5	Col 6
Propylene 115-07-1	VOC	NA	NA	3.0E+00 C	NA
Styrene 100-42-5	VOC	Changes in red blood cells, reduced red blood cell counts and hemoglobin; increased liver weight, liver, kidney and stomach lesions; neurological effects (e.g., increased reaction time, decreased memory, concentration); possibly carcinogenic in humans (IARC 2002)	NA	1.0E+00 I	2.00E+00
Tetrachloroethene (PCE) 127-18-4	VOC	Respiratory system, liver, kidney, and central nervous system effects; potential carcinogen (liver) (CDC 2010)	5.9E-09 C	2.7E-01 A	237 (8 hr MEG)
Toluene 108-88-3	VOC	Increased liver and kidney weight, nephropathy, neurological effects (e.g., vision impairment, increased performance time)	NA	5.0E+00 I	4.60E+00
Trichloroethene (TCE) 79-01-6	VOC	Respiratory system, heart, liver, kidney, and central nervous system effects; potential carcinogen (liver, kidney, non-Hodgkin's lymphoma) (IARC 2010)	2.0E-09	NA	270 (8 hr MEG)
Trichlorofluoromethane 75-69-4	VOC	Accelerated mortality, elevated incidences of pleuritis and pericarditis	NA	7.0E-01 H	4.80E+00
Xylenes (Total) 1330-20-7	VOC	Decreased body weight, increased mortality, eye and respiratory tract irritation, neurological effects (e.g., impaired learning and motor performance) (ATSDR 2007; EPA 2011)	NA	1.0E-01 I	1.06E+01
Antimony 7440-36-0	Metals	Cardiovascular effects (altered electrocardiograph and myocardial damage), respiratory effects (focal and interstitial fibrosis, edema), limited evidence of reproductive and developmental toxicity	NA	NA	NA
Lead 7439-92-1	Metals	Probable carcinogen, lung and kidney tumors, neurotoxicity, developmental delays, hypertension, impaired hearing acuity, impaired hemoglobin synthesis, and male reproductive impairment	NA	NA	1.5E-03
Zinc 7440-66-6	Metals	Inflammatory response in the lungs (ATSDR 2005)	NA	NA	7.2E-01
PM	PM	Cardiovascular and respiratory effects, disease, and mortality; reproductive and developmental effects; lung cancer (EPA 2009)	NA	NA	4.0E-02 for $PM_{2.5}$ 7.0E-02 for PM_{10}

continued

TABLE 5-1 Continued

Chemical Name and CAS Number	Class	Long-Term Health Effects[a]	Inhalation Unit Risk (mg/m^3)	Reference Concentration (mg/m^3)[c]	1-yr Air MEG (mg/m^3)[d]
2,3,7,8-Tetrachlorodibenzo-p-dioxin (TCDD) 1746-01-6	Dioxin	Likely carcinogen; cardiovascular effects, diabetes, immunologic response, altered neurobehavior, reproductive and developmental effects, birth defects; changes to the endocrine system; wasting syndrome (EPA 2003a)	3.8E01 C	4.0E-08 C	1.1E-08

NOTE: MEG = military exposure guideline; NA = not available.

[a]Health effects by any route of exposure as described in EPA IRIS chemical profiles are presented unless otherwise noted in text (EPA 2011); effects are based primarily on animal experiments.

[b]Carcinogenicity determined by EPA IRIS as follows: Likely to be Carcinogenic to Humans : available tumor effects and other key data are adequate to demonstrate carcinogenic potential to humans, but does not reach the weight-of-evidence for the descriptor "carcinogenic to humans"; Suggestive Evidence of Carcinogenic Potential: evidence from human or animal data is suggestive of carcinogenicity, which raises a concern for carcinogenic effects but is judged not sufficient for a stronger conclusion; Inadequate Information to Assess Carcinogenic Potential: available data are judged inadequate to perform an assessment; Not Likely to be Carcinogenic to Humans: available data are considered robust for deciding that there is no basis for human hazard concern (EPA 2005).

[c]Data from EPA IRIS; EPA HEAST; EPA PPRTV; CalEPA; or ATSDR as noted : I = IRIS; P = PPRTV; A = ATSDR; C = Cal EPA; X = PPRTV; H = HEAST. (U.S. Environmental Protection Agency Regions 3, 6, and 9. Regional Screening Levels for Chemical Contaminants at Superfund Sites; http://www.epa.gov/reg3hwmd/risk/human/rb-concentration_table/index.htm; accessed October 18, 2010).

[d]From: USACHPPM Technical Guide 230, Chemical Exposure Guidelines for Deployed Military Personnel, Appendix C (http://www-nehc.med.navy.mil/downloads/prevmed/TG230.pdf; accessed October 17, 2010).

weaknesses of epidemiologic studies, concluded that there is limited or suggestive evidence of an association between TCDD exposure and hypertension and ischemic heart disease (IOM 2008). Humblet et al. (2008) conducted an exhaustive literature review to also evaluate the evidence for an association between TCDD exposure and cardiovascular disease morbidity or mortality. Acknowledging that confounders were not adjusted for in every study, they found a consistent association between TCDD exposure and increased risk of ischemic heart disease and to a lesser extent an increased risk of all-cardiovascular disease (Humblet et al. 2008).

These reviews specifically evaluated TCDD, but laboratory animal data show that other 2,3,7,8 polychlorinated dibenzo dioxin and furan congeners act by similar mechanisms. It is thus generally presumed that 2,3,7,8-chlorinated congeners have similar effects; this has been demonstrated for some congeners and some health effects. These congeners are generally assessed by comparing their toxicity to that of TCDD using Toxicity Equivalence Factors (TEFs), the values of which have been estimated most recently by a World Health Organization committee (van den Berg et al. 2006).

Particulate Matter

PM air pollution includes smoke, fumes, soot, and other anthropogenic by-products, primarily from combustion sources, as well as particles from natural sources (dust, pollen, sea salt, forest fires) (Dockery 2009). PM measured at JBB would have included windblown dust and sand as well as combustion by-products. Although PM_{10} contains coarse particles with aerodynamic diameters between 2.5 mm and 10 mm, it also includes fine particles with aerodynamic diameters below 2.5 mm. PM_{10} does not include coarse particles with aerodynamic diameters greater than 10 mm, which in desert climes often constitute most of the airborne particle mass. Most of the long-term health risks that have been associated with PM_{10} in ambient air is now attributed to the $PM_{2.5}$, part of PM_{10} (EPA 2009).

Dust

The Department of Defense (DoD) has been monitoring PM in the Middle East since 2001 with the beginning of the war in Afghanistan. In 2006, the DoD initiated its Enhanced Particulate Matter Surveillance Program (EPMSP) at 15 sites in the Middle East including Djibouti, Afghanistan, Qatar, United Arab Emirates, Iraq, and Kuwait to measure total suspended particles, PM_{10}, and $PM_{2.5}$. The EPMSP stated that PM dusts identified in the region were most likely from three sources—geologic dust, burn pits, and metal sources such as lead smelting and manufacturing—but the actual sources of the air pollution were not identified by the DoD (NRC 2010). The EPMSP reported that $PM_{2.5}$ concentrations exceeded the CHPPM 1-year Military Exposure Guideline (MEG) concentration of 15 µg/m^3. Other studies have also found that U.S. National Ambient Air Quality Standards (NAAQS) for PM are exceeded in Iraq and Afghanistan (Cahill 2011).

A U.S. Navy researcher has found that dust collected in Iraq and Kuwait contains high concentrations of fine PM as well as chromium, nickel, aluminum, arsenic, and other metals; biological agents such as bacteria, viruses, and fungi were also detected (Lyles et al. 2011).

Two epidemiologic studies conducted by CHPPM and the Navy failed to find an association between exposure to ambient PM and respiratory or cardiovascular outcomes in military personnel stationed at bases with burn pits, but these studies had substantial limitations including inadequate statistical power and short follow-up (AFHSC et al. 2010). The NRC review of the EPMSP found that exposure to ambient air pollution in the Middle East could plausibly be associated with chronic health effects but further research was needed to match air monitoring with deployment of military personnel and persistent health effects (NRC 2010).

It has been suggested that high concentrations of PM from crustal sources may pose different risks for cardiovascular and respiratory effects than does PM from anthropogenic sources; nonetheless, studies have shown associations between windblown dust from the Mongolian desert and increased cardiac and respiratory morbidity in Taiwan and Korea (NRC 2010). Particular health outcomes were increased hospital admissions for COPD, cardiovascular disease, congestive heart failure, asthma, and pneumonia among others (although none of the associations was statistically significant). Studies of coarse particle dust in North America, however, have not shown such health effect associations (NRC 2010).

The EPA has established a NAAQS for fine particles of 35 µg/m³ averaged over 24 hours and 15 µg/m³ averaged over 1 year. For PM_{10}, the NAAQS is 150 µg/m³ for 24 hours and there is no annual NAAQS for coarse particles because of a lack of long-term effects associated with these particles. The EPA also found a causal relationship between long-term exposure to $PM_{2.5}$ and cardiovascular effects and mortality, and a likely causal relationship between exposure and respiratory effects. There was suggestive evidence of a causal association between long-term exposure to $PM_{2.5}$ and reproductive and developmental effects, as well as cancer, mutagenicity, and genotoxicity (EPA 2009).

Combustion-Related PM

A large database of epidemiologic literature on the health effects of exposure to combustion-related PM has documented increased cardiovascular and respiratory morbidity and mortality in the United States and internationally. In these studies, PM is characterized by its aerodynamic diameter; the most commonly studied particle sizes cutoffs are PM_{10} and $PM_{2.5}$. The American Heart Association reviews of the epidemiologic literature on ambient PM and cardiovascular disease found strong evidence that short-term (hours to weeks) and long-term (months to years) exposure to ambient PM increases risk for cardiovascular disease-related mortality and ischemic heart disease (Brook et al. 2004; Brook and Rajagopalan 2010). There is strong evidence that short-term PM exposure increases risk for cardiovascular hospitalizations and moderate evidence for increased risk for heart failure and ischemic stroke (Brook and Rajagopalan 2010).

Long-term exposure to $PM_{2.5}$ has been associated with increased cardiopulmonary and lung cancer mortality (Pope et al. 2002). Types of respiratory morbidity associated with PM exposure include increased respiratory symptoms such as cough and sneeze; increased susceptibility to infection; and exacerbation of asthma and COPD (Kelly and Fussell 2011).

Other PM Constituents

The EPMSP attempted to measure the elemental composition of the PM, including about 40 elements in the analyses (NRC 2010). The EPMSP report indicated that the average concentrations of the metals and other individual elements in the air at JBB were not likely to present a health hazard. The highest reported elemental concentrations were for soil-forming elements such as potassium, magnesium, aluminum, iron, calcium, silicon, and sulfur. Only three metals—lead, antimony, and zinc—found in all PM fractions at JBB, were reported at concentrations above the claimed analytic method detection limit (see Chapter 4). Health effects for these latter three metals are summarized in Table 5-1.

HEALTH EFFECTS OF OTHER AIR POLLUTANTS DETECTED AT JBB

Table 5-1 summarizes the long-term health effects associated with exposure to the 47 air pollutants detected at JBB plus the four additional VOCs selected above. Although the route of exposure for most of the health effects reported in the table is inhalation, effects from ingestion and dermal contact are also reported if appropriate. The table is organized by chemical class with PAHs presented first, followed by VOCs and metals, with PM and dioxins at the end. Sufficiently high exposure to these air pollutants as single chemicals has been associated with a wide variety of health effects (generally based on animal studies) from functional changes to organ damage and cancer.

Carcinogens Detected at JBB

As is usual in most air sampling efforts, a number of carcinogens were detected during the JBB air sampling campaigns, including 1 known carcinogen (benzene), 13 probably carcinogens, and 1 possible human carcinogen. Health effects for these carcinogens are given in Table 5-1. One probably human carcinogen, 1,3-butadiene, was included in the list because while it was not detected in the air sampling at JBB, burn barrel experiments (see Chapter 4) indicate that it is a likely combustion product from the burning of household waste. Types or sites of cancers or neoplastic changes in test animals associated with one or more of these air pollutants include kidney, leukemia, liver, lung, lymphoma, mammary, ovary, salivary gland, skin, and stomach (see Table 5-1).

Noncancer Health Effects

A wide range of noncancer health effects has been observed primarily in animals following exposures by various routes to the air pollutants detected at JBB, including eye and throat irritation, organ weight changes, histopathologic changes (e.g., lesions, hyperplasias), inflammation, and reduced or impaired function. These effects were found in many organs and organ systems including adrenal gland, blood, lungs, liver, kidney, stomach, spleen, and cardiovascular, respiratory, reproductive and central nervous systems. Increased or accelerated mortality was observed following exposure to trichlorofluoromethane (NCI 1978) and xylene (ATSDR 2007). Reproductive toxicity—for example, ovarian and testicular atrophy and decreased weight gain in rat dams—was observed following exposure to 2-butanone, benzene, and 1,3-butadiene. Developmental toxicity—for example, skeletal variations and decreased fetal weight—has been observed following exposure to 2-butanone, and 1,3-butadiene. Neurological and central nervous system effects include reduced nerve conduction velocity (acetone) and impaired learning and memory functions (acetone, carbon disulfide, styrene, toluene).

CUMULATIVE RISK CONSIDERATIONS

The screening risk assessments performed by the Army (Taylor et al. 2008; USAPHC 2010) indicate that the measured concentrations of all the individual chemicals are unlikely to cause health effects as they were below concentrations associated with an acceptable risk of health effects. However, health risks may be greater due to multiple pollutants, cumulative risk. Cumulative risk assessment can be used to characterize the effects of multiple exposures based on the dose and known effects of each pollutant. Since dose is dependent on several external (exposure magnitude, duration, frequency, and route) and internal (absorption, distribution, metabolism, and excretion) factors the committee could not conduct a formal cumulative risk assessment with available data, see Box 5-1.

A simple way to evaluate possible effects of multiple contaminants or cumulative exposures is to consider target organs or specific effects that are shared by many of the chemicals of concern and dose (EPA 1989, 2000, 2003b). These effects may be more likely to occur when exposure is to multiple pollutants all individually capable of causing them, and more likely to occur as the cumulative dose of the pollutants increases. For example, although JBB personnel may be exposed to many pollutants that are liver toxicants, the dose of any specific liver toxicant is

---

BOX 5-1
Factors Determining Exposure and Dose

Magnitude—Toxicant concentration in contaminated medium
Duration—Length of time exposed (minutes, hours, days, lifetime)
Frequency—How often exposure occurs (e.g., daily, seasonally)
Route—Inhalation, ingestion, or dermal exposure
Absorption—Intake and uptake processes allowing substances to cross external and internal membranes and enter the bloodstream
Distribution—Transport of absorbed material from point of absorption to tissues and fluids
Metabolism—Biochemical processes by which chemicals are subjected to change by living organisms
Excretion—Elimination of toxicants and other substances from the body

not great enough to impart an intolerable level of risk. However, exposure to multiple chemicals, all affecting liver function but not present at high doses individually, may cause liver damage collectively. To address the concerns of effects of multiple contaminants, the 2010 screening assessment (USAPHC 2010) attempted to screen for target organ effects, but accounted only for the primary target organ for each chemical (USAPHC 2010). The data summarized in Table 5-1 takes account of multiple potential target organs for each chemical, and includes 15 known, probable, or possible carcinogens affecting multiple tumor types or sites. There are also numerous pollutants with common target organs and systems, including central nervous system (15 pollutants), liver (15 pollutants), lungs/respiratory (11 pollutants), kidney (12 pollutants), blood (7 pollutants), heart or vascular (7 pollutants), reproductive (3 pollutants), developmental (5 pollutants), eye (8 pollutants), skin (5 pollutants) and spleen (1 pollutant). The presence of multiple pollutants in the air at JBB, many capable of causing similar health effects, suggests that there is likely an increased risk for such health effects from exposure to the ambient air. These organs or organ systems potentially affected by multiple chemicals constitute reasonable targets for epidemiologic monitoring.

CONCLUSIONS

The health effects of dioxin and PM are well characterized on the basis of toxicological, clinical, and observational epidemiologic studies. The health effects from exposure to dioxin and dioxin-like compounds include cancer, diabetes, and other endocrine system effects, immunologic response, neurological effects, reproductive and developmental effects, birth defects, and wasting syndrome. The health effects of PM exposure include lung cancer mortality and other types of cardiovascular and respiratory morbidity and mortality.

The data on the other pollutants reviewed here were compiled from a variety of summary sources that reviewed animal studies and less common epidemiologic investigations. The exposure conditions in many of these studies bear little resemblance to those experienced by military personnel at JBB or other locations. This hazard assessment identifies potential health effects that are biologically plausible but not definitively associated with human exposures in particular conditions. The data reviewed indicate that the potential long-term health effects associated with burn pit emissions could include any of the health effects discussed in this chapter. Numerous chemicals are associated with health effects in specific organs or organ systems. Health effects associated with five or more detected chemicals include:

- Neurological, reduced CNS function;
- Liver toxicity, reduced liver function;
- Certain cancers (stomach, respiratory, skin, and leukemia, among others);
- Respiratory toxicity and morbidity;
- Kidney toxicity and reduced kidney function;
- Blood effects (anemia, changes in various blood cell types);
- Cardiovascular toxicity and morbidity; and
- Reproductive and developmental toxicity.

Evaluating the health effects associated with a particular pollutant yields hypotheses about potential health effects that may occur upon exposure to pollutant mixtures. These hypotheses can be investigated in two ways:

- Review existing epidemiologic literature on health outcomes associated with exposures to burn pit emissions (for example, recent studies on military populations) or to combustion sources similar to burn pit emissions (for example, firefighters and others) (see Chapter 6); or
- Conduct new epidemiologic investigations (see Chapter 8).

Chapter 5 has summarized health effects data from studies of exposures to particulate matter, dioxins, and 56 air pollutants detected in sampling at JBB. These data on single pollutant exposures have limited predictive value for deployed personnel at JBB or other burn pit locations because those personnel are known to have been exposed to complex combinations of the many pollutants identified in Chapter 4, but the exact combinations of pollutants, their magnitude, and the duration of exposure are unknown. Therefore, the findings presented in this chapter are preliminary at best. The committee's recommendations on the potential long-term health effects of exposure to air pollutants at JBB, including burn pit emissions, will incorporate these data as well as the epidemiologic data review in the next chapter.

REFERENCES

AFHSC (U.S. Armed Forces Health Surveillance Center), Naval Health Research Center, and U.S. Army Public Health Command. 2010. *Epidemiological studies of health outcomes among troops deployed to burn pit sites*. Silver Spring, MD: Defense Technical Information Center.

ATSDR (Agency for Toxic Substances and Disease Registry). 1998. Toxicological profile for jet fuels JP-5 and JP-8. *ToxProfiles*. http://www.atsdr.cdc.gov/ToxProfiles/TP.asp?id=773&tid=150 (accessed July 9, 2010).

ATSDR. 2005. Toxicological profile for zinc. *ToxProfiles*. http://www.atsdr.cdc.gov/ToxProfiles/tp60-c3.pdf (accessed July 9, 2010).

ATSDR. 2007. Toxicological profile for xylenes. *ToxProfiles*. http://www.atsdr.cdc.gov/ToxProfiles/tp71-c3.pdf (accessed July 9, 2010).

Becher, H., D. Flesch-Janys, T. Kauppinen, M. Kogevinas, K. Steindorf, A. Manz, and J. Wahrendorf. 1996. Cancer mortality in German male workers exposed to phenoxy herbicides and dioxins. *Cancer Causes & Control* 7(3):312-321.

Becher, H., K. Steindorf, and D. Flesch-Janys. 1998. Quantitative cancer risk assessment for dioxins using an occupational cohort. *Environmental Health Perspectives* 106(Suppl 2):663-670.

Boman, B. C., A. B. Forsberg, and B. G. Jarvholm. 2003. Adverse health effects from ambient air pollution in relation to residential wood combustion in modern society. *Scandinavian Journal of Work, Environment and Health* 29(4):251-260.

Boman, B. C., A. B. Forsberg, and T. Sandstrom. 2006. Shedding new light on wood smoke: A risk factor for respiratory health. *European Respiratory Journal* 27(3):446-447.

Brook, R. D., and S. Rajagopalan. 2010. Particulate matter air pollution and atherosclerosis. *Current Atherosclerosis Reports* 12(5):291-300.

Brook, R. D., B. Franklin, W. Cascio, Y. Hong, G. Howard, M. Lipsett, R. Luepker, M. Mittleman, J. Samet, S. C. Smith, Jr., I. Tager. 2004. Air pollution and cardiovascular disease: A statement for healthcare professionals from the Expert Panel on Population and Prevention Science of the American Heart Association. *Circulation* 109(21):2655-2671.

Cahill, C. F. 2011. UAF's unique applied research capabilities: Unmanned aircraft and airborne contaminants. Fairbanks, AK: University of Alaska Fairbanks.

CDC (Centers for Disease Control and Prevention). 2010. *NIOSH pocket guide to chemical hazards*. November 22, 2010. http://www.cdc.gov/niosh/npg/ (accessed February 23, 2011).

Dockery, D. W. 2009. Health effects of particulate air pollution. *Annals of Epidemiology* 19(4):257-63.

EPA (U.S. Environmental Protection Agency). 1989. *Risk assessment guidance for Superfund: Vol I. Human health evaluation manual (Part A)*. EPA/540/1-89/002. Washington, DC: Environmental Protection Agency.

EPA. 2000. *Supplementary guidance for conducting health risk assessment of chemical mixtures: Risk Assessment Forum Technical Panel*. EPA/630/R-00/002. Washington, DC: U.S. Environmental Protection Agency.

EPA. 2003a. Exposure and human health reassessment of 2,3,7,8-tetrachlorodibenzo-p-dioxin (TCDD) and related compounds National Academy of Sciences (NAS) review draft. EPA/600/P-00/001Cb. Washington, DC: U.S. Environmental Protection Agency. December.

EPA. 2003b. *Framework for cumulative risk assessment*. EPA/630/P-02/001F. Washington, DC: U.S. Environmental Protection Agency.

EPA. 2005. *Fact sheet: EPA's guidelines for carcinogen risk assessment*. EPA/630/P-03/001F. http://epa.gov/cancerguidelines/cancer-guidelines-factsheet.htm (accessed October 12, 2010).

EPA. 2009. *Integrated science assessment for particulate matter (final report)*. EPA/600/R-08/139F. Research Triangle Park, NC: U.S. Environmental Protection Agency.

EPA. 2011. *Integrated risk information system*. http://www.epa.gov/IRIS/ (accessed March 8, 2011).

Fingerhut, M. A., W. E. Halperin, D. A. Marlow, L. A. Piacitelli, P. A. Honchar, M. H. Sweeney, A. L. Greife, P. A. Dill, K. Steenland, and A. J. Suruda. 1991. Cancer mortality in workers exposed to 2,3,7,8-tetrachlorodibenzo-p-dioxin. *New England Journal of Medicine* 324(4):212-218.

Hooiveld, M., D. Heederik, and H. B. de Mesquita. 1996. Preliminary results of the second follow-up of a Dutch cohort of workers occupationally exposed to phenoxy herbicides, chlorophenols and contaminants. *Organohalogen Compounds* 20:185-189.

Humblet, O., L. Birnbaum, E. Rimm, M. A. Mittleman, and R. Hauser. 2008. Dioxins and cardiovascular disease mortality. *Environmental Health Perspectives* 116(11):1443-1448.

IARC (International Agency for Research on Cancer). 1997. Polychlorinated dibenzo-para-dioxins and polychlorinated dibenzofurans. *IARC Monographs on the Evaluation of Carcinogenic Risks to Humans* 69. Lyon, France: IARC Press.

IARC. 2002. Some traditional herbal medicines, some mycotoxins, napthalene and styrene. *IARC Monographs on the Evaluation of Carcinogenic Risks to Humans* 82. Lyon, France: IARC Press.

IARC. 2004. Tobacco smoke and involuntary smoking. *IARC Monographs on the Evaluation of Carcinogenic Risks to Humans* 83. Lyon, France: IARC Press.

IARC. 2010. Some non-heterocyclic polycyclic aromatic hydrocarbons and some related exposures. *IARC Monographs on the Evaluation of Carcinogenic Risks to Humans* 92. Lyon, France: IARC Press.

IOM (Institute of Medicine). 2005. *Gulf War and health: Volume 3. Fuels, combustion products, and propellants.* Washington, DC: The National Academies Press.

IOM. 2008. *Veterans and Agent Orange: Update 2008.* Washington, DC: The National Academies Press.

IOM. 2009. *Combating tobacco use in military and veteran populations.* Washington, DC: The National Academies Press.

IOM. 2010. *Secondhand smoke exposure and acute coronary events.* Washington, DC: The National Academies Press.

Kelly, F. J., and J. C. Fussell. 2011. Air pollution and airway disease. *Clinical and Experimental Allergy* 41(8):1059-1071.

Lemieux, P. M., B. K. Gullett, C. C. Lutes, C. K. Winterrowd, and D. L. Winters. 2003. Variables affecting emissions of PCDD/Fs from uncontrolled combustion of household waste in barrels. *Journal of the Air & Waste Management Association* 53(5):523-531.

Lemieux, P. M., C. C. Lutes, and D. A. Santoianni. 2004. Emissions of organic air toxics from open burning: A comprehensive review. *Progress in Energy and Combustion Science* 30(1):1-32.

Lyles, M. B., H. L. Fredrickson, A. J. Bednar, H. B. Fannin, D. W. Griffin, and T. M. Sobecki. 2011. Medical geology: Dust exposure and potential health risks in the Middle East. *34th International Symposium on Remote Sensing of Environment.* Sydney, AU. April 10-15, 2011. http://www.isprs.org/proceedings/2011/ISRSE-34/ (accessed May 3, 2011).

Mandal, P. K. 2005. Dioxin: A review of its environmental effects and its aryl hydrocarbon receptor biology. *Journal of Comparative Physiology B-Biochemical Systemic and Environmental Physiology* 175(4):221-230.

Naeher, L. P., M. Brauer, M. Lipsett, J. T. Zelikoff, C. D. Simpson, J. Q. Koenig, and K. R. Smith. 2007. Woodsmoke health effects: A review. *Inhalation Toxicology* 19(1):67-106.

NCI (National Cancer Institute). 1978. *Bioassay of trichlorofluoromethane for possible carcinogenicity.* CAS No. 75-69-4. NCI-CG-TR 106. Bethesda, MD: National Institutes of Health.

NLM (U.S. National Library of Medicine). 2011. *Hazardous substances data bank.* http://toxnet.nlm.nih.gov/cgi-bin/sis/htmlgen?HSDB (accessed March 9, 2011).

NRC (National Research Council). 1983. *Risk assessment in the federal government, managing the process.* Washington, DC: National Academy Press.

NRC. 2006. *Health risks from dioxin and related compounds: Evaluation of the EPA reassessment.* Washington, DC: The National Academies Press.

NRC. 2009. *Science and decisions: Advancing risk assessment.* Washington, DC: The National Academies Press.

NRC. 2010. *Review of the Department of Defense Enhanced Particulate Matter Surveillance Program report.* Washington, DC: The National Academies Press.

NTP (National Toxiology Program). 1986. Toxicology and carcinogeneisis studies of xylenes (mixed) (60% m-xylene, 14% p-xylene, 9% 0-xylene, and 17% ethylbenzene) (CAS No. 1330-20-7) in F344/N rats and B6C3F1 mice (gavage studies). NTP TR 327. Research Triangle Park, NC: National Institutes of Health.

Ott, M. G., and A. Zober. 1996. Cause-specific mortality and cancer incidence among employees exposed to 2,3,7,8-TCDD after a 1953 reactor accident. *Occupational & Environmental Medicine* 53(9):606-612.

Pope, C. A., 3rd, R. T. Burnett, M. J. Thun, E. E. Calle, D. Krewski, K. Ito, and G. D. Thurston. 2002. Lung cancer, cardiopulmonary mortality, and long-term exposure to fine particulate air pollution. *Journal of the American Medical Association* 287(9):1132-1141.

Steenland, K., P. Bertazzi, A. Baccarelli, and M. Kogevinas. 2004. Dioxin revisited: Developments since the 1997 IARC classification of dioxin as a human carcinogen. *Environmental Health Perspectives* 112(13):1265-1268.

Steenland, K., J. Deddens, and L. Piacitelli. 2001. Risk assessment for 2,3,7,8-tetrachlorodibenzo-p-dioxin (TCDD) based on an epidemiologic study. *American Journal of Epidemiology* 154(5):451-458.

Taylor, G., V. Rush, A. Peck, and J. A. Vietas. 2008. *Screening health risk assessment burn pit exposures Balad Air Base, Iraq and addendum report.* IOH-RS-BR-TR-2008-0001/USACHPPM 47-MA-08PV-08. Brooks City-Base, TX: Air Force Institute for Operational Health and U.S. Army Center for Health Promotion and Preventative Medicine.

U.S. Surgeon General. 2004. *The health consequences of smoking: A report of the Surgeon General*. Washington, DC: U.S. Department of Health and Human Services, Public Health Service.

USAPHC (U.S. Army Public Health Command). 2010. *Screening health risk assessments, Joint Base Balad, Iraq, 11 May–19 June 2009*. Aberdeen Proving Ground, MD: U.S. Army Center for Health Promotion and Preventive Medicine.

Van den Berg M., L. S. Birnbaum, M. Denison, M. De Vito, W. Farland, M. Feeley, H. Fiedler, H. Hakansson, A. Hanberg, L. Haws, M. Rose, S. Safe, D. Schrenk, C. Tohyama, A. Tritscher, J. Tuomisto, M. Tysklind, N. Walker, and R. E. Peterson. 2006. The 2005 World Health Organization reevaluation of human and mammalian toxic equivalency factors for dioxins and dioxin-like compounds. *Toxicological Sciences* 093(2):223-241.

White, S. S., and L. S. Birnbaum. 2009. An overview of the effects of dioxins and dioxin-like compounds on vertebrates, as documented in human and ecological epidemiology. *Journal of Environmental Science and Health Part C-Environmental Carcinogenesis & Ecotoxicology Reviews* 27(4):197-211.

6

Health Effects Associated with Combustion Products

This chapter examines the epidemiologic studies on the relationship between adverse long-term health outcomes and exposure to combustion products thought by the committee to be comparable to those emitted from the military burn pits in Iraq and Afghanistan. The media has described disheartening stories of returning Iraq and Afghanistan veterans with unusual and often multiple medical problems, anecdotally associated with exposure to smoke from burn pits. Stories published in the *New York Times* and *Washington Post* (both on August 6, 2010) describe individuals with disabling respiratory diseases, reports of constrictive bronchiolitis (an unusual lung disease), leukemia, and other cancers, and a claim of increased rates of asthma, all suggested to be linked to exposures to burn pits (Glod 2010; Risen 2010). However, such anecdotal reports do not demonstrate causality or even association; the committee looked instead to the epidemiologic literature on the exposed populations, and on populations similarly exposed. This chapter describes the committee's approach to reviewing the literature, the main uncertainties and limitations associated with the studies, summarizes the Department of Defense's (DoD's) report of several epidemiologic studies particular to health effects and burn pit exposure, summarizes the available literature by health outcome, and presents the committee's conclusions.

APPROACH TO THE EPIDEMIOLOGIC LITERATURE

Chapters 4 and 5 identified specific contaminants present at Joint Base Balad (JBB) and outlined their potential human health effects. These health effects are, in many cases, based on animal studies, and usually only pertain to exposure to a single chemical, not mixtures of chemicals from burning materials. Thus, the committee thought it necessary to evaluate human health effects to the complex mixture of chemicals resulting from combustion. The committee began by reviewing health studies on military personnel exposed to burn pits in Iraq and Afghanistan. As discussed in Chapter 3, however, there are few such studies available. Thus, the committee decided to approach its review of the health effects stemming from exposure to burn pits by identifying populations that were considered to be the most similar to military personnel with regard to exposures to burn pits or other sources of combustion products. The committee then conducted an extensive literature search for epidemiologic studies on long-term health outcomes seen in those populations. Pertinent studies were reviewed and classified as "key" or "supporting" based on their quality and relevance to the task. These key and supporting studies formed the basis of the committee's weight-of-the-evidence approach and its conclusions on the degree of association demonstrated between exposure to combustion products and long-term health outcomes. In the following sections, the committee discusses the selection and characteristics of the surrogate populations, the methods used for the literature searches, the criteria to distinguish key and supporting studies, and the categories of association on which the committee's conclusions were based.

Populations of Interest

The committee reviewed and evaluated the epidemiologic literature for studies on populations with inhalation exposure to chemical mixtures that were considered to be similar to burn pit emissions, that is, mixtures formed by combustion of a variety of materials and waste in occupational and environmental exposure settings. Two occupational groups were identified as most likely to have comparable exposures: firefighters, including those with exposures to wildland and chemical fires, and incinerator workers. Firefighters are exposed to highly complex chemical mixtures (McGregor 2005; IARC 2010). The short intermittent spikes in exposure for firefighters are likely to differ from the long-term, chronic exposures to burn pit emissions on military bases; nevertheless, studies on firefighters are useful as the best available representation of exposures to mixtures of combustion products.

The waste disposed in burn pits is described by the DoD as municipal waste (Taylor et al. 2008). Therefore, occupational exposures to emissions from municipal incinerators were considered to be another surrogate for exposure of military personnel to burn pit emissions. Furthermore, because military personnel at JBB and other burn pit locations not only work on the base but also live there, the committee considered the literature on the health effects seen in residents living near municipal incinerators to be of interest. The committee acknowledges that exposures to emissions from municipal waste incinerators likely differ from exposures to burn pit emissions, and the value of these studies in understanding the health effects of burn pit exposures is limited.

Studies of military personnel exposed to smoke from oil-well fires in Kuwait during the 1990–1991 Gulf War were also considered. Assessments of health effects among Gulf War veterans are particularly useful because of the common background exposures (for example, dusty environment, vehicle exhaust, munitions) and personnel characteristics (for example, underlying health, exposure to stressors, general demographics) shared by those deployed to Operation Enduring Freedom (OEF) in Afghanistan and Operation Iraqi Freedom (OIF) in Iraq.

The committee acknowledges that its ability to compare exposures among the populations of interest is restricted by the unknown degree to which exposures of varying intensity, duration (short-term, intermittent exposure to combustion products for firefighters; chronic exposures for incinerator workers and those living near incinerators; short-term exposure to oil-well–fire smoke in Kuwait), and composition can be extrapolated to the burn pit exposure of military personnel at JBB and elsewhere. Military personnel at JBB might have been exposed for a few days or up to 12 months as they lived and worked on the base whereas firefighters and incinerator workers might experience occupational exposure for many years, and residents near incinerators might be subject to a lifetime of exposure to pollutants. The committee recognized that JBB personnel may have had days of high exposures when smoke and emissions from the burn pits spread across the camp, but on other days there may have been less smoke, and the overall level of emissions was unknown. Exposure to burn pit emissions via ingestion and dermal contact is an even greater unknown as no sampling of surfaces and soil was conducted. Exposure to combustion products among all groups is likely affected by time-dependent changes in engineering or other controls. For example, some firefighter studies cited in this chapter were conducted before the use of self-contained breathing apparatus and other protective gear was common, while other studies assessed firefighters using protective gear that minimized exposure. The same is true for occupational and environment exposure to incinerator emissions; engineering controls to minimize hazardous emissions have been implemented over time. At JBB, the composition and volume of the burn pit changed as practices to separate waste and the use of incinerators were implemented. Since the composition of combustion products varies greatly depending on burn characteristics and fuel, and little is known about specific exposures to the burn pits at JBB and elsewhere, the committee was unable to directly compare constituents and concentrations of the pollutants that military personnel at JBB and the surrogate populations were exposed to, nor was it able to compare the duration and frequency of these exposures.

Furthermore, all the groups considered in this chapter experience a variety of additional exposures independent of their exposure to combustion products from burn pits, fires, or municipal incinerator emissions. These additional exposures include emissions from diesel engines (aircraft, vehicular, and machinery), kerosene heaters, and other environmental stressors such as combat exposure, job-related stress, and other environmental pollutants such as dust storms. The committee focused on health effects related to combustion products as related to burn pits and did not attempt to assess health effects from these other exposures.

The committee did not consider studies of health effects reported for first responders to the World Trade Center attacks in 2001 because the composition of the smoke and emissions from this event are substantially different from combustion emissions, particularly those expected from burn pits.

Methods

Extensive searches of the scientific literature published after 1980 were conducted using two major biomedical databases: MEDLINE and EMBASE. The literature search for long-term health effects among firefighters retrieved over 400 studies, including studies of structural fires, wildland or forest fires, and chemical fires. The titles and abstracts of those studies were reviewed and studies that did not appear to be immediately relevant were deleted from the database. Deleted studies included those not linked to inhalation exposure (such as studies of job-related stress); studies that had fewer than 10 participants; studies of acute or short-term health effects only (unless considered relevant to long-term effects); studies of exposures to uranium and other types of radiation; studies reporting behavioral or psychiatric outcomes; or studies that assessed DNA or other cellular damage. The literature search for incineration workers and residents of nearby communities also returned over 400 studies. In this case, rejected studies included those that were not linked to inhalation exposures; studies that reported acute or short-term health effects only; studies that were modeling studies of emissions or that focused on children, genetics, or DNA damage; and studies of ambient air pollution. Studies that characterized emissions from incinerators but not their health effects, or that focused on waste management, were also rejected.

The committee adopted a policy of using only published papers that had undergone peer review as the basis of its conclusions. An exception was made for the epidemiologic studies conducted by the DoD to assess health effects in military personnel exposed to burn pits; theses studies are summarized below and discussed after the peer reviewed epidemiologic studies for each health outcome. Since epidemiologic studies of Gulf War veterans have been described previously by other Institute of Medicine (IOM) committees, most recently in Volume 8 of the *Gulf War and Health* series (IOM 2010), this committee relied on those assessments supplemented with a review of more recent publications.

Key and Supporting Studies

After the removal of the extraneous studies, the full text of the remaining articles and reports were retrieved. For each health outcome, committee members reviewed the studies most closely related to their area of expertise, to determine whether the criteria for a key or supporting study were met. Consistent with previous IOM reports (IOM 2010), to be designated as key, a study had to be published in a peer-reviewed journal, present information about the putative exposure and specific health outcomes, demonstrate rigorous methods, include methodological details adequate to allow a thorough assessment, and use an appropriate control or reference group. A supporting study typically had methodological limitations, such as lack of a rigorous or well-defined diagnostic method or a lack of an appropriate control group. The committee as a group reviewed the key and supporting studies identified by the committee members responsible for each health outcome. The strengths and limitations of each study and its categorization as key or supporting were discussed in plenary session and a consensus reached on its contribution to the evidence base for each category of association for each health outcome. After having reviewed all the studies in detail, the committee based its conclusions primarily on key studies. Supporting studies are included as part of the committee's analysis because they provide information that might modify confidence in the conclusions based on key studies, but they carry less weight than key studies. The committee considered the DoD epidemiologic studies as supporting literature when making conclusions about associations between exposure to combustion products and health outcomes.

Categories of Association

For its conclusions, the committee agreed to use the categories of association that have been established and used by previous IOM committees, such as those that prepared the *Veterans and Agent Orange* reports and the *Gulf War and Health* series. These categories of association have been accepted for more than a decade by Congress, the Department of Veterans Affairs (VA) and the DoD, researchers, and veterans' groups.

The five categories describe different levels of association;[1] the validity of an association is likely to vary to the extent to which common sources of spurious associations could be ruled out as the reason for the observed association. Accordingly, the criteria for each category express a degree of confidence based on the extent to which sources of error and bias were reduced. The committee discussed the evidence and reached consensus on the categorization of the evidence for each health outcome in this chapter. The committee used the following categories:

- Sufficient Evidence of a Causal Relationship: Evidence is sufficient to conclude that a causal relationship exists between exposure to combustion products and a health outcome in humans. The evidence fulfills the criteria for sufficient evidence of a causal association and satisfies several of the criteria used to assess causality: strength of association, dose–response relationship, consistency of association, temporal relationship, specificity of association, and biologic plausibility.
- Sufficient Evidence of an Association: Evidence is sufficient to conclude that there is a positive association. That is, a positive association has been observed between exposure to combustion products and a health outcome in human studies in which bias and confounding could be ruled out with reasonable confidence.
- Limited/Suggestive Evidence of an Association: Evidence is suggestive of an association between exposure to combustion products and a health outcome in humans, but this is limited because chance, bias, and confounding could not be ruled out with confidence.
- Inadequate/Insufficient Evidence to Determine Whether an Association Does or Does Not Exist: The available studies are of insufficient quality, consistency, or statistical power to permit a conclusion regarding the presence or absence of an association between exposure to combustion products and a health outcome in humans.
- Limited/Suggestive Evidence of No Association: There are several adequate studies, covering the full range of levels of exposure that humans are known to encounter, that are mutually consistent in not showing a positive association between exposure to combustion products and a health outcome. A conclusion of no association is inevitably limited to the conditions, levels of exposure, and length of observation covered by the available studies. In addition, the possibility of a very small increase in risk at the levels of exposure studied can never be excluded.

UNCERTAINTY AND LIMITATIONS OF THE STUDIES

The studies cited in this chapter have limitations and uncertainties, some common to epidemiologic studies in general, and some specific to studies of occupational populations. These limitations and uncertainties include

- Healthy worker effect—Studies of firefighters are likely to be biased downward when the comparison group is the general population, that is, risk estimates might reflect a lower risk than really exists because firefighters must meet physical health standards for employment, and must remain healthy to continue working. Thus, firefighters might have a better health status than members of the general population of the same sex and age.
- Exposure misclassification—None of the studies cited in this chapter have actual measures of inhalation to combustion products. Without measured individual exposure information, an individual might be assigned the wrong level of exposure thus masking the association between effect and exposure. Most studies use employment as a firefighter (yes/no) as the only measure of exposure, although a few studies used additional

---

[1]The following categories of association are excerpted from *Gulf War and Health: Volume 1* (IOM 2000).

measures to better define exposures, such as the number of years employed or number of fires attended. Studies of communities in the vicinity of an incinerator rely on distance from the incinerator as the best surrogate of residential exposure, using either classification into concentric rings around the site or modeled exposure estimates. Ecological study designs are limited to using information on residential history, which can lead to exposure misclassification. Not all studies report the type of waste being burned, the age or technological practices of the incinerator, or adherence to government regulations, all of which affect the amount and constituents of the emissions. Furthermore, communities might be affected by other pollution sources, such as local industry, so that exposure to an environmental contaminant cannot be wholly attributed to the incinerator.

- Lack of information on confounders—Most of the studies do not adjust for potential confounders such as tobacco smoking and alcohol consumption. The use of tobacco products, particularly cigarettes, has been causally associated with long-term adverse health effects (U.S. Surgeon General 1964). Military personnel have a greater prevalence of tobacco use than civilians, particularly when deployed where smoking rates might be as high as 50% (IOM 2009). Tobacco smoke contains many environmental contaminants, including particulate matter (PM), acrolein, polyaromatic hydrocarbons (PAHs), benzene, and metals. The 2004 U.S. Surgeon General's report associated tobacco smoke with cancer, cardiovascular disease (CVD), pulmonary disease, gastrointestinal disease, and reproductive effects (U.S. Surgeon General 2004). Even exposure to secondhand smoke can result in long-term health effects, in particular, an increased risk for lung cancer (IARC 2004) and CVD (IOM 2010).
- Limited statistical power—Small sample size in many of the studies prevents the detection of associations for the less common health outcomes such as rare cancers.
- Disease misclassification—Many of the studies in this chapter investigate mortality based on the cause of death listed on death certificates. The validity of these mortality studies is dependent on the accuracy of the reported cause of death.
- Publication bias—It is likely that the evidence base for some health outcomes is affected by publication bias, that is, results that are positive or statistically significant are more likely to be published than null results.

The variability of the studies' results and methods makes comparison across them difficult. Variables include different criteria for reference populations, lack of adjustment for confounding factors, and different statistical methods. In addition, there is uncertainty regarding the degree of similarity between the exposures to combustion products in the studies and exposure to the emissions from the burn pits. Despite these limitations, the studies reviewed in this chapter provide useful evidence on the potential health effects that might be associated with exposure to burn pits. They also highlight the many challenges inherent in the conduct of any epidemiologic study of exposure to complex mixtures.

HEALTH OUTCOMES

Health outcomes were investigated by organ system. The committee drew conclusions for the following health outcomes: respiratory, circulatory, neurologic, reproductive and developmental effects, and cancer. In addition, the committee examined the literature on other outcomes such as the autoimmune disorders systemic lupus erythematosus and rheumatoid arthritis, and on chronic multisymptom illness because these health outcomes were evaluated in the DoD epidemiologic studies of OEF/OIF military personnel deployed to sites with burn pits (AFHSC et al. 2010).

Results from the epidemiologic literature are reported here for each health outcome and organized by population (firefighters, incinerator workers and surrounding communities, and Gulf War veterans exposed to smoke from oil-well fires). Little information is available on health effects linked directly to burn pit exposure; however, two epidemiologic studies conducted by the Armed Forces Surveillance Center, Naval Health Research Center (NHRC), and U.S. Army Public Health Command (AFHSC et al. 2010) on health outcomes among OEF/OIF troops deployed to bases with burn pits are considered. A brief discussion of the key and supporting studies and a conclusion are provided for each health outcome. The committee assigns a category of association for each health outcome after a summary of the results.

For readers interested in more details for the key and supporting studies, descriptions of the study design, population, exposures, outcomes measured, adjustments, and limitations are given in tabular format in Appendix C in alphabetical order by study author rather than by health outcome to avoid duplication of studies reporting on multiple outcomes.

DoD Epidemiologic Investigations

In May 2010, the Armed Forces Health Surveillance Center (AFHSC), the Naval Health Research Center (NHRC), and the U.S. Army Public Health Command (APHC) released a report on five epidemiologic studies of military personnel deployed to burn pit sites in Iraq (AFHSC et al. 2010). In the studies, exposure was defined as deployment to a site with an active burn pit as individual exposure data were not available. The AFHSC retrospective cohort study compared the incidence rates of various diseases, among deployed (two locations with burn pits, two locations without burn pits, and Korea) and never deployed cohorts. The cohorts consisted of Army and Air Force personnel deployed between January 1, 2005, and June 30, 2007, to one of four U.S. Central Command (CENTCOM) bases or to the Republic of Korea. CENTCOM bases were Joint Base Balad (JBB) and Camp Taji in Iraq, both of which had burn pits, and Camp Buehring and Camp Arifan in Kuwait which did not have burn pits. Active-duty personnel who were located within a 3-mile radius of a burn pit were included in the exposed groups. There were 15,908 personnel who served at JBB; 2,522 personnel at Taji; and 51,299 personnel at bases without burn pits. Military personnel were included in the study if they served at least 31 days at a base by the end of their deployment in order to capture any health effects resulting from being at the base. Camps in the Republic of Korea had no burn pits but were subject to urban air pollution and PM from the surrounding desert. The comparison group consisted of 237,714 active-duty personnel stationed in the United States and not previously deployed. All individuals were followed from their return from deployment, or April 15, 2006, and censored at the earliest occurrence of a diagnosis of interest, separation from active service, start of subsequent deployment or change of station, or the end of the 36-month follow-up period. The analysis adjusted for age, race, grade, and service. The report included several different investigations: (1) incidence rates of respiratory conditions, circulatory disease, CVD, sleep apnea, and ill-defined conditions for deployed personnel versus nondeployed personnel; (2) responses to post-deployment health surveys were compared between deployed personnel at sites with or without active burn pits; and (3) medical encounters for respiratory outcomes were compared for deployed personnel at sites with or without burn pits. The investigation of medical encounters while deployed is not discussed in this chapter because such encounters were considered to relate to acute, rather than long-term, health effects (AFHSC et al. 2010).

This AFHSC study looked only at health effects occurring within 36 months after return from a site with an active burn pit. Follow-up was not long enough to detect diseases with long latency, such as cancer. There was no adjustment for confounders such as smoking. This study had a large population and was able to capture individuals' health status using electronic medical records. The DoD concluded that, based on in-theater reports of respiratory problems and the high proportion of Air Force personnel reporting exposure to burn pits at JBB, acute respiratory effects are of concern, and possible long-term health effects are not discussed in the report (AFHSC et al. 2010).

The DoD report also contains four NHRC studies that looked at the personnel stationed at the same bases as for the AFHSC study but the NHRC also included a third base in Iraq with a burn pit, Camp Speicher. Exposure was based on being located within a 5-mile radius of a documented burn pit. The first study assessed birth outcomes in infants of military personnel exposed before or during pregnancy to burn pits and is discussed in the section of this chapter on reproductive and developmental outcomes. The second study looked at respiratory health of military personnel who had been exposed to burn pits and were participants in the Millennium Cohort Study; this study is discussed in the section of this chapter on respiratory outcomes. The third and fourth studies, also of participants of the Millennium Cohort Study, focused on service members who had been exposed to burn pits and their risk of having chronic multisymptom illness (CMI), or of having physician-diagnosed lupus or rheumatoid arthritis, respectively. The Millennium Cohort examined by NHRC consisted of more than 27,000 personnel deployed in support of OEF/OIF and included over 3,000 participants considered exposed, with at least one deployment within a 5-mile radius of a documented burn pit. Exposed participants were compared with participants who were deployed to locations without burn pits. The Millennium Cohort is considered to represent U.S. military personnel, with reliable self-reported information obtained prior to enrollment and unaffected by subsequent health status.

CMI was defined by the reporting of at least two symptoms from the following categories: general fatigue, mood and cognition, and musculoskeletal. CMI was not significantly associated (p = 0.16) with being deployed within a 5-mile radius of a burn pit, cumulative exposure to a burn pit overall, or being deployed to JBB or Camps Taji or Speicher, when adjusted for sex, birth year, education, service component, service branch, pay grade, smoking status, alcohol-related problems, mental health symptoms, and baseline CMI status. However, cumulative exposure to a burn pit for more than 210 days showed a slight increase in risk for CMI (OR 1.22, 95% CI 1.04–1.44) after adjustment.

There was no association between a new diagnosis of lupus and being within 5 miles of a burn pit, cumulative exposure, or being deployed to Camp Taji or Camp Speicher. There was, however, a significant increase in the likelihood of a lupus diagnosis for those deployed to JBB (OR 3.52, 95% CI 1.59–7.79) compared with those deployed to locations without burn pits. For rheumatoid arthritis, there was no association with deployment to a burn pit location, cumulative days exposed, or camp site. One exception was an increase in rheumatoid arthritis diagnoses for those exposed to burn pits for 132–211 days (OR 2.03, 95% CI 1.18–3.49), although exposure for more than 211 days was not significant. Electronic medical records were used to confirm 33% of self-reported lupus cases and 17% of self-reported rheumatoid arthritis cases among active-duty personnel diagnosed while in the military. Among verified cases, no association between lupus or rheumatoid arthritis and exposure to burn pits was found (AFHSC et al. 2010).

There are several limitations to using the Millennium Cohort data. Confirmation of self-reported medical issues is difficult for participants who are not active-duty allowing for disease misclassification. The rare occurrences of bronchitis, emphysema, lupus, and rheumatoid arthritis and short average follow-up (2.8 years) compromise the precision of the risk estimates (AFHSC et al. 2010). Exposure misclassification is also possible as individual exposure information was not available.

The committee categorized these DoD studies as supporting due to the short period of follow-up (36 months), ecologic nature, lack of information on other hazardous environmental exposures common in the context of desert and war (for example, smoking, diesel exhaust, kerosene heaters, PM, local and regional pollution). However, as the only studies of health effects and burn pit exposure, they are uniquely valuable to the current assessment and provide the first indications of adverse health effects resulting from exposure to burn pits. The lack of additional studies and further followup to corroborate or refute the DoD's reported findings prevent the committee from being able to make decisions about the strength of an association between burn pits and the reported health outcomes.

DISEASES OF THE RESPIRATORY SYSTEM

Environmental conditions experienced by military personnel in Iraq and Afghanistan might cause respiratory effects from exposure to windblown dust, local combustion sources, and volatile evaporative emissions. The local combustion sources include burn pits or other waste incinerators, compression ignition vehicles, aircraft engines, diesel electric generators, and local industry and households. Although local contributions of wood smoke might be minimal (there is mention in the sampling field notes of a local brush fire causing smoky conditions), exposure to wood smoke from burn pits would have been likely (burning of materials such as shipping pallets). Asthma, bronchitis, chronic obstructive pulmonary disease (COPD), and respiratory symptoms have been reported to occur more frequently than expected among Gulf War veterans (IOM 2010). A retrospective case-control study found a higher risk of new-onset asthma (OR 1.58, 95% CI 1.18–2.11) among military personnel who served in OEF/OIF, compared with age- and sex-matched personnel deployed in the United States (Szema 2010). Personnel exposed to combustion products from burn pits might be at increased risk of respiratory diseases as some chemicals released by the burning of waste (as described in Chapters 4 and 5), such as acrolein and PM, are known to cause respiratory effects (see Chapter 5).

This section focuses on long-term, nonmalignant adverse respiratory conditions resulting from exposures to combustion products that are considered to be similar to burn pit emissions. First, respiratory outcomes (assessments of respiratory disease and pulmonary function indicative of potential disease) related to occupational exposures of firefighters (including firefighters involved in structural, wildland, and chemical fires) are considered. Next, respiratory outcomes from exposure to incinerators, both for workers and surrounding communities are discussed. Lastly, respiratory outcomes for veterans exposed to oil-well fires in the 1990–1991 Gulf War and the preliminary data available for veterans from OEF/OIF will be examined. Details of the studies presented in this section are found in Appendix C.

Respiratory Disease in Firefighters

The committee recognizes that firefighter exposures may be very different depending on the type of fire. Structural firefighters primarily work to extinguish fires on anthropogenic objects—for example, buildings, furniture, manufactured items—whereas wildland firefighters are exposed to combustion products from the burning of the natural environment, that is, forests and grasslands. Firefighters working to extinguish chemical fires might be exposed to a wide variety of combustion products as well as the unburned chemical(s) itself.

Key Studies

No key studies of respiratory diseases in firefighters were identified by the committee.

Supporting Studies

Fifteen studies were considered to be supportive. Of the 13 studies reporting on mortality, all reported no significant increase and even reductions in mortality from respiratory causes (Eliopulos et al. 1984; Feuer and Rosenman 1986; Vena and Fiedler 1987; Heyer et al. 1990; Rosenstock et al. 1990; Beaumont et al. 1991; Grimes et al. 1991; Demers et al. 1992a, 1995; Guidotti 1993; Aronson et al. 1994; Baris et al. 2001; Ma et al. 2005). In the largest study, Ma et al. (2005), found firefighters to have significantly lower mortality rates for all respiratory causes, and for pneumonia specifically, compared with the general Florida population. However, the 13 studies had one or more limitations that precluded their categorization as key studies including too few deaths for meaningful statistical analyses, exposure assessments that were dichotomous (employed as a firefighter, Y/N) or absent, and failure to consider tobacco smoking or the healthy worker effect.

There is concern that environmental exposures could contribute to respiratory diseases of unknown cause such as sarcoidosis. Sarcoidosis is a systemic disease characterized by granulomatous inflammation, most often involving lymph nodes and the lung, but also involving the eyes, skin, liver, heart, and central nervous system. Prezant et al. (1999) studied the annual incidence and point prevalence of biopsy-proven sarcoidosis in New York City firefighters and emergency medical personnel between 1985 and 1988. The average annual incidence among firefighters was 12.9 cases per 100,000 firefighters and the point prevalence in 1998 was 222 cases per 100,000 firefighters. The majority of those with sarcoidosis (23 of 25) had minimal impairment as assessed by radiograph (x-ray and CT scan) and pulmonary function testing. These data suggest an association between firefighting and sarcoidosis, but confirmatory studies are lacking and causation was not demonstrated.

There are few studies on the health consequences of fighting chemical fires. One longitudinal follow-up study of firefighters exposed to a 1985 fire burning polyvinyl chloride found that exposed firefighters had significantly more respiratory symptoms (cough, wheeze, shortness of breath, chest pains) at both 5–6 weeks and 22 months postexposure than unexposed firefighters with the exception of wheezing at 22 months (Markowitz 1989). Among exposed firefighters, the incidence of respiratory symptoms showed a decreasing trend over time and respiratory scores between the two time points were well correlated. The findings were similar among current, past, and never smokers. After 22 months, 12 of 64 (18%) of exposed firefighters had been diagnosed with asthma or bronchitis by a physician whereas none of the 22 controls had these diagnoses. This study is limited to a single heavy exposure to a specific set of chemicals.

Pulmonary Function in Firefighters

Pulmonary function tests are frequently used to diagnose respiratory diseases such as asthma, bronchitis, emphysema, or fibrosis. Measurements include spirometry (the flow rate and volume of air that is inhaled or exhaled), diffusion capacity (how well oxygen moves from the lungs into the blood), and lung volumes (the total amount of air in the lungs). Testing might be used to evaluate shortness of breath, diagnose disease, and track disease progression or effects of treatments/medicines (Medline Plus 2011). Pulmonary function effects can be observed even in the absence of clinical symptoms or disease.

Key Studies

Sparrow et al. (1982) conducted a longitudinal study of pulmonary function in 168 male firefighters who were participants in the larger Normative Aging Study of 2,280 male military veterans that began in 1963 in Boston. Spirometric measurements, as well as a survey of smoking habits and respiratory symptoms, were collected at 5-year intervals. The control group was a non-firefighting population from the same study. The authors found a significantly greater loss of forced vital capacity (FVC) and forced expiratory volume in one second (FEV_1) in the firefighters even after adjusting for smoking, age, height, and initial pulmonary function level ($p < 0.05$). Few respiratory symptoms and diseases were reported during follow-up, with no differences between firefighters and controls.

Peters et al. (1974) studied pulmonary function in 1,430 Boston firefighters. Repeat pulmonary function tests and questionnaires collecting self-reported respiratory symptoms and smoking habits were completed from 1970 to 1972. Pulmonary function declined in the entire cohort (FVC annual loss of 77 mL, FEV1 of 68 mL) and was significantly associated with frequency of exposure to fires ($p < 0.01$). Decreases could not be explained by the effects of age, smoking, or race. Additional follow-up of 1,146 firefighters through 1974 showed that decreased pulmonary function and association with numbers of fires fought was maintained (Musk et al. 1978). However, a further follow-up of this cohort for a total of 6 years through 1976, found smaller declines in FVC and FEV1, no correlation with exposure, and no significant difference from healthy nonsmoking non-firefighters (Musk et al. 1982). This change was attributed to increased use of protective respiratory equipment.

Supporting Studies

Supporting studies of pulmonary function show mixed results. Several studies report no decrease in pulmonary function for firefighters while other studies indicate increases in respiratory symptoms. Decreased pulmonary function was reported for structural firefighters (Unger et al. 1980; Tepper et al. 1991) and for forest firefighters (Liu et al. 1992; Serra et al. 1996; Betchley et al. 1997).

An examination of respiratory function among 128 firefighters and 88 controls in Zagreb, Croatia, found significantly higher rates ($p < 0.01$) of respiratory symptoms (dyspnea, nasal catarrh, sinusitis, and hoarseness) and decreased pulmonary function in firefighters compared to controls (Mustajbegovic et al. 2001). The authors found these chronic respiratory symptoms and decreases in pulmonary function to be associated with duration of employment and smoking. Young et al. (1980) conducted a cross-sectional study of respiratory disease and pulmonary function among 193 firefighters in New South Wales, Australia. The authors found no increased respiratory problems attributable to fire exposure and concluded that "the major combustion products responsible for respiratory damage were self-administered, arising from burning tobacco rather than from burning buildings." Miedinger et al. (2007) examined respiratory symptoms, atopy, and bronchial hyperreactivity in 101 professional firefighters compared with 735 local men in Basel, Switzerland. Firefighters had better FEV_1, FVC (significant), and FEV_1/FVC values than controls, although they also had elevated rates of respiratory symptoms, atopy, and bronchial hyperreactivity (OR 2.24, 95% CI 1.12–4.48). Douglas et al. (1985) examined the effect of firefighting on the pulmonary function of 1,006 London firefighters over 1 year. Lower than expected pulmonary function was not associated with exposure, based on years of employment and self-reported exposure to severe smoke events, except for a nonsignificant decrease among firemen who had worked for more than 20 years. This analysis adjusted for smoking.

Horsfield et al. (1988a,b) conducted two longitudinal studies of respiratory health in 96 West Sussex firefighters and 69 local nonsmoking men assessed every six months for 2 years and annually thereafter over the course of 4 years. Respiratory symptoms increased at a faster rate for firemen, regardless of smoking status, compared with the nonsmoking controls, leading the authors to conclude "these results suggest that being affected by smoke and fumes at work may be a cause of long-term symptoms in firemen" (Horsfield et al. 1988b). However, the control group had a greater decrease in pulmonary function and spirometric measurements than the firemen, leading the authors to further conclude that "these results show no evidence of chronic lung damage in West Sussex firemen" (Horsfield et al. 1988a). They attributed these findings not only to a healthy-worker selection bias but to the increasing use of protective breathing apparatus by these firemen.

Unger (1980) investigated the acute and chronic effects of a severe smoke exposure event on the pulmonary function of 30 firefighters sent to a Houston-area hospital after a single fire. Spirometric data and a survey of self-reported symptoms were collected immediately, after 6 weeks, and again after 18 months. Significant decreases in FVC ($p < 0.01$) and FEV_1 ($p < 0.05$) were observed compared to matched controls. Tepper et al. (1991) evaluated pulmonary function changes after 6 to 10 years in male Baltimore firefighters (n = 632) in a longitudinal cohort study that adjusted for age, smoking, blood type, and weight. Firefighters who did not wear respirators and those who were exposed to ammonia were both found to have 1.7 times the rate of decrease in FEV_1 of unexposed controls. Active firefighting was associated with 2.5 times ($p < 0.05$) the rate of decrease compared with individuals no longer working as firefighters.

Three studies investigated forest firefighters. Two examined cross-season differences in respiratory function. Liu et al. (1992) conducted a longitudinal study of 63 seasonal and full-time wildland firefighters in Northern California and Montana, pre- and post firefighting season in 1989. The authors found a postseasonal loss in lung function (0.15 L FEV_1) and an increase in airway responsiveness (significant mean declines in FVC, $FEV_1$1, and FEF_{25-75}) compared with preseason values after controlling for smoking. Betchley et al. (1997) observed significant decreases ($p < 0.05$) in cross-season spirometry values for 53 forest firefighters based on questionnaires and testing before and after the 1992 firefighting season. Mean individual decreases were 0.033 L for FVC_1, 0.104 L for FEV_1, and 0.275 L/sec for FEF_{25-75}. The authors also reported significant decreases in pulmonary function across shifts for the 72 individuals assessed for cross-shift differences. Results were not affected by smoking, recent colds, lung conditions, allergies, or other potential confounders with the exception of those relying on wood to heat their homes. The third study of forest firefighters by Serra et al. (1996), examined pulmonary function in 92 Sardinian forest firefighters compared with 51 local police officers. The firefighters had significant decreases in FEV_1, FVC, FEF_{75}, FEV_1/FVC, FEF_{50}, FEF_{25}, but the decreases were not correlated with length of service or number of fires extinguished after adjusting for age, height, smoking status, and pack-years. No difference in permeability of alveolar-capillary barrier was observed.

Respiratory Disease in Incinerator Workers

No studies of occupational respiratory disease among incinerator workers were identified by the committee.

Pulmonary Function in Incinerator Workers

Key Studies

No key studies of pulmonary function in incinerator workers were identified.

Supporting Studies

Three supporting studies of pulmonary function in incinerator workers were considered by the committee. Bresnitz et al. (1992) conducted a cross-sectional study of 89 male incinerator workers in Philadelphia, Pennsylvania. The study included environmental monitoring, physical examinations of all study participants, analysis of biological samples, and pulmonary function tests. The prevalence of pulmonary function patterns were similar in high and low exposed groups, after adjusting for smoking status. The OR for small airway obstruction in the high versus low exposed group was 1.19 (95% CI = 0.45–3.16). Changes in pulmonary function were related only to smoking status. Conversely, two other studies report decreased pulmonary function among incinerator workers after adjusting for smoking. Charbotel et al. (2005) noted significantly reduced pulmonary function from predicted values in the third year of monitoring—FEF_{50} ($p = 0.04$), FEF_{25-75} ($p = 0.02$) and FEF_{25-75}/FVC ($p = 0.01$)—among 83 workers exposed to incinerator emissions compared with 76 unexposed workers, indicating possible obstructive disorders for the exposed workers. After adjusting for history of allergy or lung disease, smoking, and location of examination, the reduction of FEF_{75} in the first year and FEF_{25-75}/FVC in the third year were linked to exposure in incinerator plants. Charbotel et al. (2005) noted that daily variation in lung function may not have been captured. A study of 102 male workers at three French urban incinerators by job type was conducted by Hours et al. (2003). Symptoms were self-reported on a survey and a physical exam was performed with blood testing and respiratory function assessment. Workers were compared with 84 water-meter assemblers, security guards, or woolen-mill workers. Daily coughing was reported more often by incinerator furnace men (OR = 6.58, 95% CI 2.18–19.85) and decreased respiratory performance was found in incinerator maintenance and effluent treatment workers ($p < 0.01$) after adjusting for smoking, age, and work location.

Respiratory Disease in Communities Near Incinerators

No studies assessing respiratory disease incidence or mortality in populations exposed to incinerator emissins were identified by the committee.

Pulmonary Function in Communities Near Incinerators

Key Studies

No key studies of pulmonary function in populations exposed to incinerator emissions were identified by the committee.

Supporting Studies

Four studies, conducted as part of the Health and Clean Air Study (Shy et al. 1995), assessed health outcomes in communities living near incinerators. Exposure was based on distance from an incinerator. No significant differences were noted between respiratory symptoms or pulmonary function and community exposure to incinerator emissions (Shy et al. 1995; Lee and Shy 1999; Hu et al. 2001; Hazucha et al. 2002). Using the Health and Clean Air Study (Shy et al. 1995) and one additional community near a commercial hazardous waste incinerator, Mohan et al. (2000) compared respiratory symptoms with four control communities matched by socioeconomic characteristics and population size. The authors found a higher prevalence of all respiratory symptoms in the one community near a hazardous waste incinerator compared with the control community ($p < 0.05$) even after controlling for perceptions of air quality.

Respiratory Disease in Gulf War Veterans Exposed to Oil-Well–Fire Smoke

Studies of respiratory outcomes related to exposure to smoke from oil-well fires were reviewed by previous IOM committees tasked with assessing long-term health effects in Gulf War veterans (IOM 2006). Those committees noted that these veteran studies are valuable for their relatively robust exposure estimates; however, the studies generally lack the temporal context to distinguish between new respiratory illness and pre-existing conditions (for example, asthma that was present before deployment versus asthma onset after exposure to oil-well fire smoke).

Gulf War and Health: Volume 4 (IOM 2006) identified three key studies,[2] all using similar methods to describe troop exposure to smoke from oil-well fires by linking troop locations and National Oceanic and Atmospheric Administration (NOAA) meteorologic information, and to assess risks of respiratory diseases (Cowan 2002; Lange 2002; Smith 2002). Cowan et al. (2002) conducted a case-control study to identify cases of physician-diagnosed asthma in a DoD registry of clinically evaluated active-duty Gulf War Veterans (n = 873) and controls without asthma (n = 2,464). Self-reported oil-well–fire smoke exposure was associated with a higher risk of asthma (OR 1.56, 95% CI 1.23–1.97). In addition, modeled cumulative oil-well–fire smoke exposure was also related to a greater risk of asthma (OR 1.08, 95% CI 1.01–1.15) and showed the greatest risk among those with greater exposures (OR 1.21, 95% CI 0.97–1.51 for the intermediate-exposure group of up to 1.0 mg-day/m^3; and OR 1.40, 95% CI 1.12–1.76 for the high-exposure group of over 1.0 mg-day/m^3). The study controlled for sex, age, race or ethnicity, rank, smoking history, and self-reported exposure. When exposure was classified as number of days with exposure at 65 µg/m^3 or greater, the risk of asthma also increased with longer exposures. Study strengths include the objective exposure assessment and the use of physician-diagnosed asthma as the basis of clinical evaluations. Limitations include the lack of pulmonary function data and specified criteria for the diagnosis of asthma, and self-selection into the DoD registry.

A study of a population-based Iowa cohort of 1,560 Gulf War veterans found no statistical association between modeled oil-well–fire exposure and the risk of asthma (Lange 2002). Five years after the war, veterans were asked about their exposures and current symptoms. Self-reported exposure to oil-well fires was associated with a greater risk of asthma and bronchitis. However, there was no statistical association between modeled exposure and the risk of asthma or bronchitis in models that controlled for sex, age, race, military rank, smoking history, military service, and level of preparedness. The authors ascribed the different results for self-reported and objective exposure measurement to recall bias. Population-based sampling, which implies that findings can be generalized to all military personnel in the Persian Gulf, is a strength; however, it is limited by poor case definitions and disease misclassification.

In a postwar hospitalization study of 405,142 active-duty Gulf War veterans, Smith et al. (2002) also examined the effect of oil-well-fire exposure. There was no association between exposure to the fires and the risk of hospitalization for asthma (relative risk [RR] 0.90, 95% CI 0.74–1.10), acute bronchitis (RR 1.09, 95% CI 0.62–1.90), or chronic bronchitis (RR 0.78, 95% CI 0.38–1.57). There was a modest nonsignificant increase in the relative risk of emphysema (RR 1.36, 95% CI 0.62–2.98). Because most adults who have asthma or chronic bronchitis are never hospitalized for the condition, the study would not be expected to have captured most cases. No information was available on tobacco-smoking or other exposures that may be related to respiratory symptoms.

Respiratory Disease in OEF/OIF Veterans Exposed to Burn Pits

The DoD report (AFHSC et al. 2010) describing several epidemiologic investigations of health outcomes among military personnel deployed to Iraq and Afghanistan included results on respiratory diseases. The investigators compared incidence rates of diseases and disorders at two military bases in Iraq with burn pits (JBB and Camp Taji) and with four comparison groups (nondeployed personnel in the United States, those deployed to two sites without burn pits in Kuwait [Camps Arifjan and Buehring], and those deployed to Korea). Those potentially exposed to burn pits at JBB or Camp Taji in Iraq consistently showed significantly lower or similar adjusted incidence rate ratios when compared with the nondeployed group. Incidence rate ratios (IRRs) were significantly

---

[2]The following text was excerpted from IOM (2006).

decreased among personnel deployed to JBB and Taji for respiratory diseases and acute respiratory infections, while estimates were significantly lower in JBB (not significant in Taji) for COPD, asthma, and sleep apnea with all risk estimates for both bases being less than one. However, personnel without exposure to burn pits at Camps Arifjan and Buehring in Kuwait, and at the Korean base had similarly reduced or nonsignificant IRRs, indicating a healthy warrior effect but no disease potentially associated with being deployed to sites with burn pits.

The DoD report also included details of a Millennium Cohort Study analysis that found no significant differences in newly diagnosed asthma, bronchitis, emphysema, or self-reported respiratory symptoms between those deployed to areas within 5 miles of burn pits and those not exposed. No increased risk was noted with increasing cumulative exposure (days) or by camp site. Analyses were adjusted for smoking status, physical activity, and other covariates measured at baseline (AFHSC et al. 2010).

Cases of bronchiolitis have been reported in the media, presumably from exposure to the Mosul sulfur fire in Iraq in 2003 (Bartoo 2010). Constrictive bronchiolitis (CB), also known as bronchiolitis obliterans, is a narrowing of the small airways (the bronchioles) in the lungs. It can be irreversible and can impair daily functioning to the extent that affected individuals are no longer fit for military duty. There are multiple causes, including rheumatoid arthritis, inhalation of toxicants, rejection of a transplanted lung, or it might be idiopathic. It can only be diagnosed by lung biopsy, an invasive procedure that requires a hospital stay, and there is no accepted treatment (King et al. 2008; Bartoo 2010).

Preliminary reports described an unexpected number of CB cases among military personnel who had lung biopsies to determine the cause of their shortness of breath (King et al. 2008; Miller 2009; Bartoo 2010). A total of 38 soldiers were diagnosed with constrictive brochiolitis among 49 who underwent lung biopsy (out of 80 soldiers who had served in Iraq and/or Afghanistan and were referred for respiratory problems). Of the cases, 87% reported exposure while deployed to dust storms, 74% to the 2003 sulfur fire in Mosul, Iraq, and 63% to incinerated waste. Compared to a group of healthy soldiers, the cases were significantly ($p < 0.001$) older; had a higher BMI; had reduced pulmonary function measures for FEV1, FVC, and capacity to diffuse carbon monoxide; and had worse results for several cardiopulmonary tests (King et al. 2011).

In response to these cases of CB, the US Army Surgeon General requested a further study by CHPPM at Fort Campbell, Kentucky (USAPHC 2010). The exploratory analysis of chronic or recurring lung disease among veterans exposed to the Mosul sulfur fire in 2003 (191 Army firefighters and 6,341 soldiers located within 50 km of the fire) compared with unexposed deployed troops found no association between exposure to the fire and CB, but the possibility of health effects could not be ruled out. Follow-up was conducted through June 2007. Comparison of morbidity among firefighters and the exposed brigade to two control populations provided mixed results. Compared to unexposed controls in the Q-west area, standardized mortality ratios (SMRs) were significantly decreased for respiratory diseases (acute respiratory infections, COPD, asthma, circulatory diseases, ill-defined conditions, signs and symptoms involving the cardiovascular system, and signs and symptoms involving the respiratory system, all SMRs less than 0.74 and $p < 0.05$, for the exposed firefighters and the exposed brigade. Compared to the general unexposed population, the exposed brigade was diagnosed with more respiratory diseases (SMR 1.13, 95% CI 1.08–1.18), acute respiratory infections (SMR 1.18, 95% CI 1.12–1.24), and signs, symptoms, and ill-defined conditions involving respiratory system and chest (SMR 1.08, 95% CI 1.0–1.17), whereas SMRs for other disease groups were not significant and significant differences were not reported between the firefighters and the general unexposed population. The authors note that demographic differences between exposed and unexposed groups might have introduced bias, especially as those exposed tended to be younger. Despite the study's limitations and potential biases, the authors reported no clear associations from this investigation but recommend follow-up and further study (USAPHC 2010).

Conclusions

Overall, the key and supporting studies showed no evidence of increased mortality from respiratory causes among firefighters. One small supporting study reported an increase in sarcoidosis prevalence (Prezant et al. 1999). There are no key studies of respiratory disease among incinerator workers or surrounding communities.

The two key studies of pulmonary function in firefighters show decreases in lung function using longitudinal analyses while controlling for important factors such as smoking (Peters et al. 1974; Sparrow et al. 1982). Supporting studies of pulmonary function among firefighters show a wide range of results, however, with only a few of them showing decreased pulmonary function in firefighters (Unger et al. 1980; Tepper et al. 1991; Liu et al. 1992; Serra et al. 1996; Betchley et al. 1997; Mustajbegovic et al. 2001). Two of three supporting studies of incinerator workers show decreased pulmonary function after a few years of follow-up; these studies adjusted for other causes of decreased function and symptoms and decreases specific to job-related exposures (Hours et al. 2003; Charbotel et al. 2005). No key, but several supporting studies describe community exposure to incinerator emissions as part of the Health and Clean Air Study; however, those studies do not indicate any increased risks of respiratory effects.

In conclusion, the committee finds that there is *inadequate/insufficient* evidence to determine whether an association exists between respiratory disease and combustion products in the populations discussed here. However, several studies that found reductions in pulmonary function among firefighters and incinerator workers provide *limited/suggestive* evidence for an association between exposure to combustion products and decreased pulmonary function in these populations.

DISEASES OF THE NERVOUS SYSTEM

Diseases of the nervous system encompass genetic, degenerative, and traumatic damage to the brain, spinal cord, and nerves. Risk factors for specific neurologic disorders include age, family history, infections, traumatic injuries, and exposure to environmental toxicants. Neurologic outcomes range from changes in cognitive function, headache, nerve or bodily pain, to epilepsy, amyotrophic lateral sclerosis (ALS), Parkinson's disease, and multiple sclerosis (MS). The committee identified only a few studies of neurologic diseases that were attributed to exposures similar to those expected from the burn pits. No studies of neurologic effects in communities surrounding incinerators were identified.

Neurologic Disease in Firefighters

Key Studies

The committee identified only one key study of neurologic outcomes in firefighters. Baris et al. (2001) conducted a retrospective cohort mortality study of neurologic effects in 7,789 structural firefighters in Philadelphia employed between 1925 and 1986 and a reference group of the general U.S. white male population. The firefighters worked for an average of 18 years, with a 26-year average follow-up. The measures of exposure included duration of employment, type of company employment, year of hire, cumulative number of fire runs, and fire runs during first 5 years as a fireman, as well as lifetime fire runs with diesel exposure. Based upon 12 firefighter deaths, no increase in deaths from nervous system diseases was observed (SMR 0.47, 95% CI 0.27–0.83).

Supporting Studies

Two studies have reported central nervous system effects (Kilburn et al. 1989; Bandaranayke et al. 1993) in firefighters exposed to polychlorinated biphenyls (PCBs) and other chemicals from specific events. Bandaranayke et al. (1993) found that 245 firefighters exposed to a chemical fire were more likely to experience nervous system effects as evidenced by poor neuropsychological test responses, but effects from chronic anxiety could not be ruled out. Kilburn et al. (1989) reported impaired memory and cognitive function among 14 firefighters exposed to PCBs, but the study quality is poor, and a reanalysis of the data suggested that most of the significant findings could be due to chance alone (Mustacchi 1991). Additionally, several cohort studies (Vena and Fiedler 1987; Beaumont et al. 1991; Grimes et al. 1991; Guidotti 1993; Tornling et al. 1994; Ma et al. 2005) also failed to find an increased risk for neurologic disease among firefighters exposed to chemical or other types of fires.

Neurological Disease in Incinerator Workers

Key Studies

No key studies of neurologic effects among those occupationally exposed to emissions from incinerators were identified by the committee.

Supporting Studies

Gustavsson (1989) found no increase in mortality from nervous system diseases among a cohort of 176 incinerator workers in Stockholm, Sweden, when compared to national and local mortality rates (SMR 1.33, 95% CI 0.03–7.39; SMR 1.32, 95% CI 0.03–7.36, respectively).

Neurologic Disease in Gulf War Veterans Exposed to Oil-Well–Fire Smoke

Several Gulf War studies evaluated the relationship between exposure to combustion products and neurobehavioral effects (Iowa Persian Gulf Study Group 1997; Proctor 1998; Unwin et al. 1999; Kang et al. 2000; Spencer et al. 2001; White et al. 2001; Wolfe et al. 2002). These studies found a positive relationship between self-reported exposure and self-reported neuropsychological, cognitive, or mood symptoms. Because combustion product exposure was self-reported, the Gulf War and Health committee considered the studies to provide only weak evidence of an effect.

Smith et al. (2002) investigated hospitalization for mental disorders and nervous system diseases by categories of exposure to smoke from oil-well fires among Gulf War veterans; relative risk ranged from 0.79 to 0.96 (all nonsignificant). Barth et al. (2009), studying mortality from ALS, MS, and Parkinson's disease, found no increased risk for any of these diseases among veterans potentially exposed to oil-well–fire smoke.

Neurologic Disease in OEF/OIF Veterans Exposed to Burn Pits

No studies of neurologic disease among OEF/OIF veterans were identified by the committee.

Conclusions

Occupational studies of firefighters and incinerator workers do not show increased rates of mortality or elevated prevalence of neurologic disease. Personnel serving in Iraq during the 1990–1991 Gulf War as firefighters and reporting exposure to oil-well–fire smoke had a greater prevalence of self-reported neurobehavioral effects than era veterans not reporting exposure; however, the increase could not be definitively associated with exposure to combustion products.

Based upon its review of the literature, the committee concludes that there was *inadequate/insufficient evidence* for an association between combustion products and nervous system disease or neurobehavioral effects in the populations studied.

DISEASES OF THE CIRCULATORY SYSTEM

Circulatory system diseases, often referred to as cardiovascular diseases (CVD), include many health outcomes pertaining to the heart and vascular tissues. Circulatory diseases include rheumatic heart disease; hypertension; ischemic heart disease; congestive heart failure; pulmonary heart disease and other forms of heart disease; cerebrovascular disease; diseases of the arteries, arterioles, capillaries, veins, and lymphatic vessels; and other circulatory system disorders.

The most common cause of circulatory disease is atherosclerosis, which is the formation of vascular lesions resulting from excessive lipid deposition and macrophage infiltration. Atherosclerosis is the main cause of cardio-

TABLE 6-1 Risk Estimates for Diseases of the Circulatory System in Firefighters and Incinerator Workers

Study	Study Type	Population		Risk Estimate (95% CI)
Diseases of the Circulatory System				
Aronson et al. 1994	K	Firefighters	SMR	0.99 (0.89–1.10)
Baris et al. 2001	K	Firefighters	SMR	1.01 (0.96–1.07)
Eliopulos et al. 1984	K	Firefighters	SMR	0.78 (0.60–1.01)
Guidotti et al. 1993	K	Firefighters	SMR	1.03 (0.88–1.21)
Heyer et al. 1990	K	Firefighters	SMR	*0.78 (0.68–0.92)*
Tornling et al. 1994	K	Firefighters	SMR	0.84 (0.71–0.98)
Vena and Fiedler 1987	K	Firefighters	SMR	0.92 (0.81–1.04)
Gustavsson 1989	K	Incinerator workers	SMR	1.06 (0.78–1.42)
Bates et al. 1987	S	Firefighters	SMR	*1.73 (1.12–2.66)*
Feuer and Rosenman 1986	S	Firefighters	PMR	1.02 (p > 0.05)
Grimes et al. 1991	S	Firefighters	SMR	*1.16 (1.10–1.32)*
Ma et al. 2005	S	Firefighters	SMR	*0.69 (0.63–0.76)*
Milham and Ossiander 2001	S	Firefighters	PMR	0.99 (p = 0.70)
Cardiovascular Disease				
Ma et al. 2005	S	Firefighters	SMR	*0.73 (0.65–0.83)*
Diseases of the Heart				
Demers et al. 1992a	K	Firefighters	SMR	*0.79 (0.72–0.87)*
Guidotti et al. 1993	K	Firefighters	SMR	1.10 (0.92–1.31)
Beaumont et al. 1991	K	Firefighters	SMR	*0.89 (0.81–0.97)*
Chronic Rheumatic Heart Disease				
Aronson et al. 1994	K	Firefighters	SMR	*0.15 (0.004–0.85)*
Milham and Ossiander 2001	S	Firefighters	PMR	0.73 (p = 0.28)
Hypertensive Diseases				
Milham and Ossiander 2001	S	Firefighters	PMR	0.82 (p = 0.20)
Ischemic Heart Disease				
Aronson et al. 1994	K	Firefighters	SMR	1.04 (0.92–1.17)
Baris et al. 2001	K	Firefighters	SMR	*1.09 (1.02–1.16)*
Beaumont et al. 1991	K	Firefighters	SMR	0.95 (0.87–1.04)
Demers et al. 1992a	K	Firefighters	SMR	*0.82 (0.74–0.90)*
Eliopulos et al. 1984	K	Firefighters	SMR	0.84 (0.60–1.14)
Guidotti et al. 1993	K	Firefighters	SMR	1.06 (0.87–1.28)
Hansen 1990	K	Firefighters	SMR	1.15 (0.74–1.78)
Tornling et al. 1994	K	Firefighters	SMR	0.98 (0.81–1.17)
Dibbs et al. 1982	K	Firefighters	RR	0.5 (0.2–1.4)
Gustavsson 1989	K	Incinerator workers	SMR	1.26 (0.87–1.76)
Burnett et al. 1994	S	Firefighters	PMR	1.01 (0.97–1.05)
Calvert et al. 1999	S	Firefighters	PMR	1.04 (0.94–1.14) white
Calvert et al. 1999	S	Firefighters	PMR	*1.69 (1.10–2.47) black*
Deschamps et al. 1995	S	Firefighters	SM	0.74 (0.20–1.90)
Sardinas et al. 1986	S	Firefighters	SMR	*1.52 (1.23–1.81)*
Sardinas et al. 1986	S	Firefighters	MOR	1.07 (0.91–1.23)
Smith et al. 2002		Gulf War	RR	*0.82 (0.68–0.99)*
Myocardial Infarction				
Aronson et al. 1994	K	Firefighters	SMR	1.07 (0.93–1.23)
Dibbs et al. 1982	K	Firefighters	RR	0.5 (0.1–1.9)
Pulmonary Embolism and Infarction				
Milham and Ossiander 2001	S	Firefighters	PMR	0.80 (p = 0.55)
Cerebrovascular Diseases				
Aronson et al. 1994	K	Firefighters	SMR	0.76 (0.55–1.03)
Baris et al. 2001	K	Firefighters	SMR	*0.83 (0.69–0.99)*
Beaumont et al. 1991	K	Firefighters	SMR	0.84 (0.67–1.03)

TABLE 6-1 Continued

Study	Study Type	Population		Risk Estimate (95% CI)
Demers et al. 1992a	K	Firefighters	SMR	0.85 (0.67–1.06)
Guidotti et al. 1993	K	Firefighters	SMR	*0.39 (0.18–0.73)*
Tornling et al. 1994	K	Firefighters	SMR	0.71 (0.44–1.09)
Vena and Fiedler 1987	K	Firefighters	SMR	0.92 (0.64–1.27)
Gustavsson 1989	K	Incinerator workers	SMR	0.45 (0.09–1.33)
Deschamps et al. 1995	S	Firefighters	SMR	1.16 (0.24–3.38)
Feuer and Rosenman 1986	S	Firefighters	PMR	0.63 ($p > 0.05$)
Grimes et al. 1991	S	Firefighters	SMR	1.13 (0.66–1.95)
Ma et al. 2005	S	Firefighters	SMR	0.80 (0.60–1.05)
Arteriosclerotic Heart Disease				
Aronson et al. 1994	K	Firefighters	SMR	1.41 (0.91–2.10)
Guidotti et al. 1993	K	Firefighters	SMR	1.50 (0.77–2.62)
Heyer et al. 1990	K	Firefighters	SMR	*0.75 (0.63–0.89)*
Vena and Fiedler 1987	K	Firefighters	SMR	0.92 (0.79–1.07)
Feuer and Rosenman 1986	S	Firefighters	PMR	1.11 ($p > 0.05$)
Grimes et al. 1991	S	Firefighters	SMR	1.09 (0.89–1.35)
Milham and Ossiander 2001	S	Firefighters	PMR	1.02 ($p = 0.39$)
Diseases of Arteries				
Milham and Ossiander 2001	S	Firefighters	PMR	0.94 ($p = 0.56$)
Aortic Aneurysm				
Aronson et al. 1994	K	Firefighters	SMR	*2.26 (1.36–3.54)*
Diseases of Veins				
Aronson et al. 1994	K	Firefighters	SMR	1.68 (0.72–3.31)
Milham and Ossiander 2001	S	Firefighters	PMR	1.10 ($p = 0.70$)
Other Circulatory Diseases				
Beaumont et al. 1991	K	Firefighters	SMR	0.88 (0.73–1.04)
Demers et al. 1992a	K	Firefighters	SMR	0.96 (0.80–1.14)
Eliopulos et al. 1984	K	Firefighters	SMR	0.65 (0.38–1.07)
Deschamps et al. 1995	S	Firefighters	SMR	0.84 (0.02–4.68)
Other Heart Diseases				
Milham and Ossiander 2001	S	Firefighters	PMR	0.98 ($p = 0.81$)
Cholesterol Abnormality				
Hu et al. 2003	S	Incinerator workers	OR	2.8 (1.0–7.9)

NOTE: Risk estimates in bold italics denote significant differences in risk. K = key study; S = supporting study.

vascular mortality associated with acute myocardial ischemia, heart failure, and stroke. Patients with myocardial ischemic injury often have arrhythmias, hypertrophy, cardiomyopathy, and heart failure. Several risk factors are known to contribute to overall circulatory disease including elevated low-density lipoprotein (LDL cholesterol), increased blood pressure, diabetes mellitus, tobacco use, poor diet, obesity, excessive alcohol use, and physical inactivity (CDC 2009).

Studies on circulatory outcomes in firefighters, incinerator workers, communities near incinerators, and Gulf War veterans are discussed below. Because of the large number of studies with risk estimates for multiple circulatory outcomes, the results have been summarized in Table 6-1 by specific circulatory effect. This section uses disease terminology as cited by study authors.

Circulatory Disease in Firefighters

Key Studies

The committee reviewed several studies reporting circulatory disease mortality and one measuring disease incidence in firefighters to assess whether occupational exposure to combustion products affects circulatory disease. A majority of those studies were designed to evaluate all-cause mortality with circulatory disease as a sub-category of evaluation.

Aronson et al. (1994) conducted a retrospective cohort mortality study of 5,414 firefighters in metropolitan Toronto with 777 deaths. SMRs for diseases of veins, cerebral vascular disease, and chronic rheumatic heart disease were not elevated compared to the general male population of Ontario after adjusting for age and calendar period. Ischemic heart disease accounted for 289 deaths (SMR 1.04, 95% CI 0.92–1.17), of which 205 were due to acute myocardial ischemia (SMR 1.07, 95% CI 0.93–1.23). A statistically significant excess of myocardial infarction was apparent at 20–24 years since first employment (SMR 1.56, 95% CI 1.03–2.25), but no trend was apparent. Arteriosclerosis accounted for 24 deaths (SMR 1.41, 95% CI 0.91–2.10), 19 of which were attributed to aortic aneurysm (SMR 2.26, 95% CI 1.36–3.54). No deaths from aortic aneurysm occurred prior to 20 years since first employment. However, at 20–30 years since first employment the SMR was 3.03 (95% CI 0.63–8.86); at 40–49 years since first employment the authors observed a statistically significant SMR of 2.95 (95% CI 1.27–5.82). The majority of deaths (17 of 19) occurred in firefighters who worked 25 years or more (SMR 2.50, 95% CI 1.46–4.00). The authors concluded that while this study provides some evidence of an association between occupation as a firefighter and risk of aortic aneurysms, the result might be due to chance because knowledge at the time did not support an occupational etiology.

Baris et al. (2001) conducted a retrospective cohort mortality study to assess cancer mortality among 7,789 firefighters in Philadelphia compared to the mortality of U.S. white men. The authors found a significant excess risk for ischemic heart disease (SMR 1.09, 95% CI 1.02–1.16). No significant association was found for circulatory diseases, and a decreased risk of cerebrovascular disease was noted (SMR 0.83, 95% CI 0.69–0.99). There were no positive exposure-response trends for any outcome across increasing years of employment or number of runs. Results were not consistent across strata when grouped by company type (engine, ladder, or both) or by date of hire. Outcomes associated with diesel exposure could not be assessed due to the small number of diesel-exposed firefighters and short follow-up.

A mortality study of San Francisco firefighters by Beaumont et al. (1991) found fewer cardiovascular deaths than expected among 3,066 firefighters employed from 1940 to 1970 and followed through 1982. There were 508 deaths from diseases of the heart and 131 other circulatory deaths. Incidence density rate ratios, indirectly standardized for age, sex, year, and race based on person-years at risk, compared the firefighters to the general U.S. population. A significant decrease in heart diseases was seen (RR 0.89, 95% CI 0.81–0.97). The risk of ischemic heart disease, other circulatory diseases, and cerebrovascular disease was decreased for firefighters, but not significantly.

The mortality of 4,546 male firefighters from three northwestern U.S. cities was assessed by Demers et al. (1992a) compared to police officers. No significantly elevated risks were noted. Rather, significantly reduced risks were found for heart disease (SMR 0.79, 95% CI 0.72–0.87; incidence density ratio [IDR] 0.86, 95% CI 0.74–1.0) and ischemic heart disease (SMR 0.82, 95% CI 0.74–0.90; IDR not significant). SMRs were reduced but not significantly for other circulatory diseases and cerebrovascular disease; however, the IDRs were 0.72 (95% CI 0.54–0.96) and 0.65 (95% CI 0.45–0.92), respectively. Neither SMR nor IDR were significant for diseases of arteries, veins, and pulmonary circulation. Incorporating a 10-year lag showed an SMR of 2.55 (95% CI 1.43–3.38) for diseases of arteries, veins, and pulmonary circulation among those with 30 or more years of exposure. Furthermore, when SMRs were stratified by duration of employment, years since first employment, and age, those in the highest categories had significantly elevated SMRs for diseases of arteries, veins, and pulmonary circulation. A previous analysis of this cohort was published by Rosenstock et al. (1990) who reported an SMR of 0.81 (95% CI 0.73–0.89) for nonmalignant circulatory diseases compared to the white U.S. male population.

Dibbs et al. (1982) conducted a cohort study of firefighters enrolled in the Normative Aging Study, a longitudinal study of aging among veterans. Four of the 171 firefighters (2.3%) developed coronary heart disease compared with 71 (4.8%) of 1,475 non-firefighters. The estimated RR for myocardial infarction was 0.5 (95% CI 0.1–1.9) and 0.5 (95% CI 0.2–1.4) for coronary heart disease, indicating a lower rate of disease among firefighters compared with non-firefighters. Risk factors including serum cholesterol, systolic and diastolic blood pressures, body mass index, and cigarette smoking were assessed. Only the percentage of smokers varied between firefighters and non-firefighters (41.5% vs. 35.7%). Relative risks were not adjusted for any covariates other than age.

Glueck et al. (1996) prospectively followed a cohort of 806 firefighters in Cincinnati, Ohio, who underwent a series of physical exams, blood tests, and cardiac tests. In the 8-year follow-up period there were 7 myocardial infarcts and 15 coronary heart disease (CHD) cases without infarction. The firefighters were significantly heavier, older, had higher systolic blood pressure, and longer follow-up than the comparison group of healthy, employed men taking part in the National Health and Nutrition Examination Survey I (NHANES I) without CHD. There were slightly fewer myocardial infarcts among the firefighters compared with the control group (1.35/1,000 person-years vs. 2.07/1,000 person-years, $p = 0.1$). It should be noted that at study entry, firefighters who sustained CHD were significantly older; had higher diastolic and systolic blood pressures; smoked more per day; had higher LDL, triglycerides, and total cholesterol; were more likely to have a family member with CHD; and had a shorter follow-up time ($p < 0.05$) than firefighters who did not develop CHD. Those who developed CHD also had a longer length of employment (22.2 years vs. 18 years, $p = 0.052$). Twenty-six firefighters reported having experienced smoke inhalation but did not develop CHD. No increase in the risk of CHD among firefighters was observed. The authors concluded that CHD risk among firefighters is due primarily to conventional and modifiable risk factors (blood pressure, cholesterol, and smoking).

Studying a subpopulation of that reported by Demers et al. (1992a), Heyer et al. (1990) studied 2,289 male firefighters in Seattle who were employed between 1945 and 1980. Compared to the U.S. white males, SMRs for circulatory system disease mortality and specifically arteriosclerotic disease were significantly decreased in firefighters (SMR 0.78, 95% CI 0.68–0.92; SMR 0.75, 95% CI 0.63–0.89, respectively). When stratified by age, time since first employment and duration of exposure, those older than 65 years, those with less than 30 years since first exposure, and firefighters with less than 30 years of exposure, were all significantly less likely to die from circulatory diseases. A subgroup of individuals surviving 30 years or more after their first exposure revealed a trend ($p < 0.10$) of increasing risk with increasing exposure for circulatory system disease; firefighters with 30 years or more of fire combat duty had a RR of 1.84 compared with those with less than 15 years of fire combat duty.

Eliopulos et al. (1984) conducted a retrospective cohort mortality study of 990 male firefighters employed by the Western Australian Fire Brigade between 1939 and 1978. With 16,876 person-years of follow-up and 116 deaths, the authors found no evidence of increased mortality from CVD when compared to deaths of West Australian males, adjusted for age and calendar time. The nonsignificant SMRs for circulatory disease (55 deaths), ischemic heart disease (39 deaths), and other circulatory diseases (16 deaths) were 0.78, 0.84, and 0.65, respectively.

Guidotti et al. (1993) conducted a retrospective cohort mortality study of 3,328 firefighters employed between 1927 and 1987 in Edmonton and Calgary, Alberta. Vital status for 96% of the cohort was ascertained, resulting in 370 deaths and 64,983 person-years. An exposure opportunity index term, reflecting estimates of the relative time spent in close proximity to fires by job classification, was applied to refine exposure data based on years of service. The authors found that mortality from heart disease was close to that expected (SMR 1.10; 95% CI 0.92–1.31) compared to the male population of Alberta when adjusted for age and calendar period. Specific causes (ischemic heart disease and arteriosclerosis) were also not significantly elevated among the firefighters and the risk of cerebrovascular disease was lower (SMR 0.39, 95% CI 0.18–0.73). No significant trends or associations were noted when stratified by exposure index or latency. The authors could not assess potential confounders and expressed concern about a lack of power when assessing trends and multiple strata.

A study of mortality among firefighters in Denmark was reported by Hansen (1990). The cohort of 866 male firefighters was identified from the 1970 nationwide census of public employees aged 15 to 69 years old. Male civil servants and public employees in the same age range (n = 47,694) employed in physically demanding jobs were used as a comparison group adjusted for age and calendar period. There were 57 deaths among firefighters and 2,383 deaths in the comparison group at 10 years of follow-up. The SMR was 1.15 (95% CI 0.74–1.71) for ischemic heart disease; however, when stratified to compare the first and last 5 years of follow-up, the SMR for the former was 1.48 (95% CI 0.74–2.65) and 0.97 (95% CI 0.52–1.66) for the latter. Other cardiovascular causes of death were not reported. While the comparison group was selected to resemble the firefighters based on physical fitness and strength, social class, geographic distribution, and stability of employment to minimize the effects of a healthy worker bias, the groups might differ in lifestyle.

In Stockholm, Tornling et al. (1994) examined the mortality of 1,116 male firefighters employed for at least a year between 1931 and 1983 and followed through 1986. Exposure was determined from 10% of all reports of all fires in Stockholm (19,000 reports examined). An exposure index was created based on number of fires fought each calendar year and use of self-contained breathing apparatus. SMRs were standardized for age, calendar year, and sex using the Stockholm population. For circulatory diseases, the SMR was 0.84 (95% CI 0.71–0.98); ischemic heart disease was 0.98 (95% CI 0.81–1.17), and cerebrovascular disease was 0.71 (95% CI 0.44–1.09). While generally not significantly elevated, SMRs for ischemic heart disease showed a tendency to increase with older age and latency. When stratified by duration of employment and number of fires, however, no trends were evident and no significant associations were noted.

Vena and Fiedler (1987) compared mortality of 1,867 white male firefighters in Buffalo, New York, with mortality of U.S. males with adjustment for age. The SMRs were 0.92 (95% CI 0.81–1.04) for all diseases of circulatory system, 0.92 (95% CI 0.79–1.07) for arteriosclerotic heart disease, and 0.92 (95% CI 0.64–1.27) for all central nervous system (CNS) vascular lesions. No increase in circulatory deaths was observed when stratified by duration of employment, calendar year of death, year of hire, or latency.

Supporting Studies

Using a subpopulation of the cohort studied by Aronson et al. (1994), Bates (1987) compared 646 firefighters employed by the Toronto Fire Department with all men living in the city of Toronto. The authors identified 21 firefighter deaths due to coronary artery disease (out of 52 total deaths) and reported no significant association between death from circulatory disease and firefighting activity or year of death. Studying ischemic heart disease in firefighters in Connecticut, Sardinas et al. (1986) reported a slightly elevated SMR of 1.52 (95% CI 1.23–1.81) and mortality odds ratio (MOR) of 1.07 (95% CI 0.91–1.23) compared to the general Connecticut population, but a significantly lower mortality among firemen when compared to policemen (MOR 0.62, 95% CI 0.56–0.68). In Washington State, (Milham and Ossiander 2001) found that age-adjusted proportional mortality ratios (PMRs) for circulatory disease, or any subcategories of circulatory disease (arteriosclerotic heart disease including coronary disease, other diseases of the heart, hypertensive disease, diseases of the arteries, diseases of the veins, or pulmonary embolism and infarction) were not elevated for firefighters and fire protection workers. Among 205 City of Honolulu firefighters, Grimes et al. (1991) found that cardiovascular mortality was slightly elevated, with a crude RR of 1.16 (95% CI 1.10–1.32). Slightly elevated but nonsignificant risks were reported for arteriosclerotic heart disease and all CNS vascular lesions; however, when stratified by ethnicity (Caucasian and Hawaiian), no associations were apparent.

Mortality of U.S. firefighters in 27 states was assessed by Burnett et al. (1994) using the National Occupational Mortality Surveillance system. There was no increase in ischemic heart disease deaths among firefighters. Another analyses of National Occupational Mortality Surveillance data investigated racial differences. Calvert et al. (1999) found that firefighters were among the 10 occupations with the highest PMRs for ischemic heart disease for black males (PMR 1.69, 95% CI 1.10–2.47) but not for white males.

In New Jersey, Feuer and Rosenman (1986) studied police and firefighters employed for at least 10 years in New Jersey and found increased risks of arteriosclerotic heart disease (PMR 1.22) compared to U.S. mortality; however, when compared with the New Jersey population, the PMR was slightly, but not significantly, elevated (PMR 1.11), and no excess was evident when compared to police officers. For circulatory deaths, no association was apparent compared to U.S. or New Jersey populations or to police officers. The investigators' decision to use three different reference groups, particularly police who are likely to be more similar to firefighters than the general New Jersey or U.S. population, mitigates bias introduced by a healthy worker effect.

Exhibiting a strong healthy worker effect, Deschamps et al. (1995) found no association between ischemic heart disease (four deaths), other circulatory diseases (one death), or cerebrovascular disease (three deaths) and occupational firefighting for 830 Parisian firefighters compared with the general French male population after 14 years of follow-up. The authors noted a number of study limitations, particularly the small number of events (32 deaths overall). The study by Ma et al. (2005) also had a healthy worker effect but had a large population of 34,796 male and 2,017 female professional firefighters in Florida. When compared to general population of Florida and adjusted for age and gender, the authors found the risks of circulatory disease were significantly reduced for male firefighters but elevated for female firefighters (deaths attributed to circulatory system causes, for men SMR 0.69, 95% CI 0.63–0.76 and for women SMR 2.49, 95% CI 1.32–4.25; atherosclerotic heart disease for male firefighters SMR 0.73, 95% CI 0.65–0.83, and for female firefighters SMR 3.85, 95% CI 1.66–7.58).

Three reviews of CVD in firefighters were identified by the committee. Haas et al. (2003) assessed six studies of coronary artery disease mortality in firefighters, all discussed above, and found no convincing evidence of an association between firefighting and coronary artery disease. According to the authors, the bias imposed by any healthy worker effect was not as strong as expected for these studies. A review of 22 studies by Guidotti (1995), 14 of which are described in this chapter, showed no presumptive risk of CVD, COPD, or aortic aneurysm among firefighters. Taking a different approach to consider the healthy worker effect, Choi (2000) evaluated 23 studies of CVD in firefighters, 17 of which are described above. Before the reassessment to factor in the healthy worker effects, 7 of the 23 studies (Eliopulos et al. 1984; Hansen 1990; Grimes et al. 1991; Guidotti 1993; Aronson et al. 1994; Burnett et al. 1994; Deschamps et al. 1995) showed an increased risk and 16 showed no increase. After the reassessment, 11studies showed an increase in risk and 12 did not. Choi concluded that after considering the healthy worker effect, there is an increased risk of death from heart disease among firefighters, but there is insufficient evidence of an increased risk of death from aortic aneurysm or of an association between firefighting and any heart disease subtype, such as MI. However, Choi could not eliminate confounding factors as the cause of the apparent increase in risks.

Circulatory Disease and Incinerator Workers

Key Studies

Gustavsson (1989) assessed mortality among 176 municipal waste incinerator workers near Stockholm, Sweden. Using national and local mortality rates to calculate SMRs, the risks for circulatory diseases, ischemic heart disease, or cerebrovascular diseases was not increased among incinerator workers. When stratified by length of employment, incinerator workers, compared with the local population, had an elevated death rate due to ischemic heart disease (SMR 1.86, $p < 0.01$) after more than 40 years after first employment, but not for shorter employment. Ischemic heart disease mortality rates were also elevated among those with more than 30 years of employment (SMR 1.67, $p < 0.05$), but not for less than 30 years.

Supporting Studies

No supporting studies of circulatory disease among municipal incinerator workers were identified by the committee.

Circulatory Disease in Communities Near Incinerators

No studies of circulatory disease among people living near incinerators were identified by the committee.

Circulatory Disease in Gulf War Veterans Exposed to Oil-Well–Fire Smoke

Among the studies of disease in Gulf War veterans, one key study examined CVD associated with exposure to oil-well fires.[3] Morbidity in the form of post-war hospitalizations at DoD medical facilities of 405,142 active-duty

---

[3]This paragraph is based on text from IOM (2010). The current committee did not review the study, rather it relied on the assessment of the prior IOM committee that examined the evidence and weighed the health effects literature on this type of exposure and cardiovascular outcomes.

military personnel deployed to the Gulf War during the time of the fires in Kuwait (February 2, 1991, to October 31, 1991) was examined by Smith et al. (2002). Hospitalizations between 1991 and 1999 for diseases of the circulatory system were evaluated. Exposure to oil-well–fire smoke was estimated by combining smoke-plume modeling data and troop unit location. Exposure was categorized into seven levels based on combinations of average daily exposure (none, 1–260 µg/m^3, or > 260 µg/m^3) and duration of exposure (1–25 days, 26–50 days, or > 50 days). Compared with military personnel with no exposure to smoke from oil-well fires, there was no increase in the incidence of hospitalization for cardiovascular disorders at any level of exposure; relative risks ranged between 0.9 and 1.2 for the different levels of exposure without a clear dose-response trend. Relative risks were significantly decreased among those exposed to 1–260 µg/m^3 for 26–50 days and > 50 days, and those exposed to > 260 µg/m^3 for > 50 days. Veterans exposed to oil-well fires had a slightly lower risk than unexposed veterans for hospitalizations for ischemic heart disease (RR 0.82, 95% CI 0.68–0.99). This study provided individual level quantifiable exposure estimates but did not include confounding behavioral and environmental exposures. A limitation of this study is that health effects needed to be severe enough for hospital admission and, thus, less severe outcomes might be missed.

Circulatory Disease in OEF/OIF Veterans Exposed to Burn Pits

The DoD determined incidence relative risks for health outcomes in military personnel deployed to sites with burn pits (JBB and Taji in Iraq), to sites without burn pits (Arifan, Buehring, and Korea), and non-deployed personnel with 3 years of follow-up. Personnel at JBB reported fewer cases of circulatory system diseases (IRR 0.94, 95% CI 0.90–0.98) and cardiovascular signs and symptoms (IRR 0.81, 95% CI 0.74–0.88) compared with nondeployed military personnel, whereas personnel stationed at Taji were no different than nondeployed personnel. However, personnel deployed to Korea and to Arifan and Buehring in Kuwait (all without any potential exposure to burn pits) had similarly reduced or not significant incidence relative risks. CVD potentially associated with being deployed to a site with burn pits was not observed (AFHSC et al. 2010).

Conclusions

Based on the 25 studies of firefighters, incinerator workers, communities around incinerators, and veterans from Gulf War and OEF/OIF, two key studies (Aronson et al. 1994; Baris et al. 2001) provide support for an increased risk of certain cardiovascular outcomes in firefighters. Most studies assessed multiple cardiovascular outcomes including all circulatory disease, heart disease, arteriosclerosis, aortic aneurysm, myocardial infarction, cerebrovascular disease, other circulatory diseases, and diseases of the arteries and veins. Increased risks for any specific cardiovascular outcome were not consistently observed in these studies. Results of these studies were affected by several limitations, particularly the inability to control for confounding and inadequate follow-up time. A healthy worker bias was introduced by the use of inappropriate control groups in many studies, such as the use of general U.S. or local populations; however, some studies attempted to use other occupational groups such as police officers to correct for this bias. Lastly, as is true for many studies in this chapter, the lack of exposure assessment makes it difficult to determine whether a possible increase in the risk of CVD can be attributed to exposure to combustion products.

Based on its review of the epidemiologic literature, the committee concludes that there is *insufficient/inadequate* evidence of an association between long-term occupational exposure to combustion products and circulatory disease in the populations studied.

ADVERSE REPRODUCTIVE AND PERINATAL OUTCOMES

Potential effects of parental exposure to combustion products on reproductive and perinatal health were of concern to the committee. Chemical exposures may affect fertility, maternal health, birth weight, infant health and development, twinning, or fetal and infant death. The committee reviewed studies on adverse reproductive and perinatal outcomes in children born to parents with three different exposure scenarios: firefighters, people living in communities near municipal incinerators, and military personnel deployed in the 1990–1991 Gulf War or OEF/OIF.

Birth Defects in Children of Male Firefighters

The following studies assess birth defects among live-born children of male firefighters. No studies assessing other adverse reproductive outcomes such as fetal death, infant death, low birth weight, and infertility among firefighter populations were identified. Sample sizes are small, and while maternal demographic characteristics are considered, there is no adjustment for maternal exposures to smoke or other toxicants during pregnancy.

Key Studies

No key studies of birth defects in children of male firefighters were identified by the committee.

Supporting Studies

Olshan et al. (1990) evaluated data on 89 malformed children born to 271 British Columbia firefighters in 1952–1973, matched against 174 normal children born to 749 policemen and normal children born to 21,929 fathers in all other occupations by month, year, and hospital of birth. Focused on heart defects, these investigators found heart defects to be strongly associated with parental firefighting as an occupation: specifically, ventricular defects (OR 2.30, 95% CI 0.77–6.92 compared with all other occupations, and OR 5.05, 95% CI 1.43–17.82 compared with police), atrial defects (OR 6.81, 95% CI 1.40–33.16 compared with all other occupations, and OR 3.82, 95% CI 1.19–12.33 compared with police); and patent ductus arteriosis (OR 1.69, 95% CI 0.45–6.39 compared with all other occupations, and OR 14.6, 95% CI 1.03–206.16 compared with police). Odds ratios were elevated but not significant for several other categories including microcephalus, cleft lip, pyloric stenosis, and urinary tract obstructive defects, with some adjustments for low birth-weight and prematurity. The investigators suggest that children of firefighters are at increased risk of ventricular and atrial septal defects, with wide confidence intervals for the other elevated risk categories.

Two subsequent case-control studies do not corroborate the cardiac findings reported by Olshan et al. (1990). Aronson et al. (1996) compared 9,340 children with congenital heart defects, all born from 1979 to 1986, with 9,340 healthy controls randomly selected from the Ontario birth certificate file, and matched by birth year, maternal age, birth order of the child, birthplace of each parent (born in Ontario vs. not born in Ontario), and marital status of the mother at the time of birth. Eleven cases and 9 controls were born to fathers employed as firefighters (OR 1.22, 95% CI 0.46–3.33). The authors concluded that there was no association between firefighting as an occupation and the kinds of cardiac defects identified by Olshan et al. (1990). Like Aronson et al. (1996), Schnitzer et al. (1995) were not able to corroborate the findings reported by Olshan et al., and did not detect increased risk of ventricular or atrial defects among children of firefighters. Schnitzer et al. (1995) examined 28 categories of major birth defects in children born to Atlanta firefighter fathers, 1968–1980, compared to fathers with malformed children in all other occupations, matched for race, year, and hospital of birth. There were increased risks for cleft palate (OR 13.3, 95% CI 4.0–44.4), other heart defects (OR 4.7, 95% 1.2–17.8), hypospadias (OR 2.6, 95% CI 1.1–6.2), and club foot (OR 2.9, 95% CI 1.4–6.0).

Two other publications reviewed those studies on reproductive outcomes among the offspring of male firefighters. McDiarmid and Agnew (1995) concluded that measures should be taken to protect male and female firefighters from reproductive toxicity of harmful chemicals, especially for pregnant firefighters given the particular sensitivity of the human fetus to carbon monoxide. In their review of epidemiologic studies on paternal occupations and birth defects, Chia and Shi (2002) cite these studies but also emphasize the studies' weaknesses and the need for more rigorous studies.

Reproductive Outcomes for Incinerator Workers

Key Studies

No key studies on reproductive outcomes and occupational exposures among incinerator workers were identified by the committee.

Supporting Studies

No supporting studies of reproductive outcomes among incinerator workers were identified by the committee.

Reproductive Outcomes for Parents Living Near Incinerators

Key Studies

No key studies pertaining to reproductive effects among people living near incinerators were reviewed by the committee.

Supporting Studies

The committee reviewed eight community-based incinerator studies in two general outcome categories: the first five studies (Jansson and Voog 1989; Cresswell et al. 2003; Cordier et al. 2004; Tango et al. 2004; Vinceti et al. 2008) examined birth defects, stillbirths, and infant death, while the last three (Lloyd et al. 1988; Williams et al. 1992; Rydhstroem 1998) looked at twinning and sex ratios.

Several studies modeled exposure to dioxin, which was generally assumed to be from the incinerators under study. In Sweden, Jansson and Voog (1989) investigated six cleft lip or cleft palate births at a single hospital during a 6-month period in 1987 to families living within a 50-km radius of an incinerator compared with local rates of birth defects for 18 Swedish boroughs. The authors concluded that modeled dioxin exposure did not explain the cluster. Taking a wider ecologic approach, no difference in local incidence rates of cleft lip or cleft palate births for each borough that installed incinerators from 1972 to 1986 was detected. The investigators suggested that no increased risk of cleft lip or cleft palate could be demonstrated in the study areas since the start of incineration practices but because of the small study group, observations could be attributed to chance. Tango et al. (2004) studied registry-based births in Japan to examine adverse reproductive effects associated with maternal residence within 10 km of one of 63 incinerators with high levels of dioxin emissions (> 80 ng-TEQ/m^3). Reporting on a wide range of reproductive outcomes (low birth-weight, very low birth-weight, neonatal deaths, all congenital malformations combined, etc.), there was no significant excess of adverse reproductive effects to women living within 2 km of the incinerators including 225,215 live births from 1997–1998. However, the ratio of observed to expected neonatal deaths declined with increasing distance from the incinerators up to 10 km (p = 0.023 for infant deaths and p = 0.047 for infant deaths with all malformations combined). Drawing cases from hospital discharge records and a regional birth defects registry for 2003–2006, Vinceti et al. (2008) found no significantly increased risks or dose-response trends for birth defects or spontaneous abortion among women living (residential cohort) or working (occupational cohort) in the vicinity of the municipal incinerator in Modena, Italy, compared to women in the remaining municipal population. Areas of high and intermediate exposure were defined using meteorological data to estimate deposition of dioxins and furans in the study area. The analysis was limited by having few cases.

Cordier et al. (2004) studied congenital anomalies in 194 exposed communities in the vicinity of 70 different incinerators in the Rhone-Alps region of southeastern France. The overall rate of congenital anomalies was not higher in exposed communities compared with 2,678 neighboring unexposed communities (RR 1.04), after adjusting for year of birth, maternal age, department, population density, and average family income (modeled using distance to incinerator, dispersion modeling, and semi-quantitative dioxin emission estimates). A few congenital anomalies showed excess risks including facial clefts (RR 1.30, 95% CI 1.06–1.59) and renal dysplasia (RR 1.55,

95% CI 1.10–2.20). The report concludes that both incinerator emissions and road traffic density might explain some of the reported excess risk. In another birth defects registry study, Cresswell et al. (2003) compared congenital anomalies in two different geographic areas defined by distance from an incinerator in the British city of Newcastle-on-Tyne that started operation in 1988. Out of 81,255 live births between 1985–1999, there were 428 cases of birth defects in families residing within 3 km and 1,080 cases residing within 3–7 km of the incinerator. Investigators reported no significant association between the number of anomalies and residential proximity to the incinerator; however, rates of congenital abnormalities became significantly higher in 1998 and 1999 (RRs of 1.56 and 2.05, respectively) compared with the earlier years. Investigators were unable to explain the results for the last 2 years of the study without information on cumulative exposure or increases in exposure over later years.

Responding to anecdotal reports of increased twinning in cattle presumably due to pollutants from two incinerators in central Scotland, Lloyd et al. (1988) compared single/twin birth rates in hospitals in areas exposed to incinerator pollutants for the years 1975 to 1983. They concluded that the increased frequency of human twinning in areas most affected by incinerators, and supported by an anecdotal increase in twinning among dairy cattle, warrants further epidemiologic study. Studying the same population, Williams et al. (1992) reported a significant excess in female births (m:f ratio of 0.87, $p < 0.05$) at a location identified as most at risk from incinerator-based air pollution, but the ratio for the three risk areas combined was 1.01. The m:f ratio for the comparison areas combined was 0.99 and the ratio for Scotland was 1.06. In another study conducted in Sweden, Rydhstroem (1998) looked at twin deliveries before and after the commissioning of 14 incinerators between 1973 and 1990. Comparing recorded twin births against the expected number standardized for maternal age and year of delivery for all Swedish parishes, the authors report no apparent increase in the number of parishes or municipalities with a relative excess of twin deliveries, and no spatial clustering after the incinerators were commissioned.

Reproductive Outcomes for Gulf War Veterans Exposed to Oil-Well–Fire Smoke

An IOM study (2005) examined the scientific literature for associations between illness and exposure to toxic agents associated with Gulf War service. This review of more than 60 studies examined a broad range of reproductive and developmental outcomes, including fertility, preterm birth, low birth-weight, birth defects, and childhood cancers. Using the same classification system as in this report, that committee concluded that the evidence was inadequate/insufficient to determine whether an association exists between parental exposure to fuels such as kerosene, gasoline, and diesel, and adverse reproductive or developmental outcomes. That committee concluded there was limited/suggestive evidence of an association between parental exposure to combustion products such as PM_{10}, NO_2, SO_2, and CO and pre-term birth and low birth-weight, but inadequate/insufficient evidence for other reproductive outcomes. The current committee did not review the studies in the 2005 report; but rather it relied on that committee's published report and any new published studies.

As described in the IOM report on the health of Gulf War veterans (IOM 2010), Verret et al. (2008), in a study deemed to be supporting in that report, evaluated the effect of Gulf War deployment on the incidence of birth defects. When compared to 10-year prevalence data from the Paris Registry of Congenital Malformations, the prevalence of major anomalies did not differ between French Gulf War veterans and the general French population, with the exception of a decrease of Down syndrome among children born to veterans (prevalence rate ratio 0.36, 95% CI 0.13–0.78). The authors also conducted a nested case-control study. No associations were observed between birth defects and self-reported exposures to oil-well fire smoke, sandstorms, chemical alarms, or pesticides. To minimize recall bias, controls were restricted to veterans with at least one symptom-related hospitalization. However, the inclusion criterion was not applied equally to the cases; thus, control selection was plausibly related to exposure, and the results of the case-control analyses were subject to selection bias.

Reproductive Outcomes in OEF/OIF Veterans Exposed to Burn Pits

One preliminary epidemiologic investigation is described in a DoD report on a wide range of health outcomes among troops deployed to burn pit sites in Iraq and Afghanistan. This investigation, conducted by the NHRC (AFHSC et al. 2010), found no increase in the risk of preterm births or birth defects during the first year of life in children born to women deployed before and during pregnancy (OR 1.19, 95% CI 0.89–1.64) out of 13,129 women deployed within a 5-mile radius of burn pits. No relationship with cumulative days of deployment or temporal proximity to conception were noted among mothers. For men deployed within a 5-mile radius of a burn pit anytime before the estimated date of conception (6,763 burn pit deployments out of 88,074 total deployed), there was no reported increase in premature births or birth defects (OR 0.98, 95% CI 0.86–1.12). However, the risk of unspecified birth defects increased if fathers were exposed more than 280 days prior to the estimated date of conception (OR 1.31, 95% CI 1.04–1.64), but no associations were noted with cumulative days of exposure.

Conclusions

The committee determined that there were no key studies of reproductive or perinatal outcomes among firefighters, incinerator workers, or communities near incinerators on which to base its conclusions. The supporting studies of birth defects in children of firefighters give mixed results, including uncorroborated reports of cardiac defects, cleft palate, hypospadias, and other defects. The studies raise questions, but they provided inadequate evidence of an association between working as a firefighter and birth defects of their children. While recognizing maternal exposure (and other maternal characteristics) as possible confounders, none of the studies appears to regard this information as critical to interpreting the results.

For children of populations living or working near incinerators, five supporting studies considered birth defects, stillbirths, and infant death. Three supporting studies examined twinning and/or sex ratios. Some elevated risks were reported, but investigators generally concluded there was no increased risk after adjustments. The committee noted several problems with these studies such as potential sources of misclassification of diagnosis, and exposure based on maternal work history, residential history, or birth registration errors. For most of the studies, maternal residence was determined at the time of birth, without data on the duration of residence in the study area relative to the pregnancy, introducing doubt about antenatal exposure in relation to windows of susceptibility for different fetal organ systems.

For children born to male and female personnel deployed in the Gulf and Iraq Wars, the committee found some evidence suggesting an association between adverse outcomes for preterm births and low birth-weight and parental exposure to combustion products, but not for birth defects or other outcomes, as detailed in the IOM reports (2005, 2010).

The committee concludes that there is *insufficient/inadequate* evidence to determine whether an association exists between parental exposure to combustion products and adverse effects in their children and other adverse reproductive and perinatal outcomes in the populations studied.

CANCER

Exposure to environmental carcinogens is estimated to cause two-thirds of cancers (National Institute of Environmental Health Sciences 2003). While lifestyle, behavioral, and genetic factors are the most important known risk factors for cancer, many environmental exposures including mineralogical substances, infectious agents, biological toxins, radiation, pesticides, particulate matter and dust, solvents, exhaust, and a variety of industrial chemicals have been linked to cancer. Of importance to veterans potentially exposed to burn pit by-products, some types of particulate matter have been linked to lung and respiratory cancers; dioxins are highly carcinogenic; and PAHs have been linked to lung, skin, and urinary cancers (NIEHS 2003).

This section describes cancer incidence and mortality studies on populations exposed to fires and combustion products. Each study is discussed below including relevant results. Given the large number of studies with risk estimates for many different cancer sites, results are also summarized by cancer site in Table 6-2 at the end of this section.

Cancer in Firefighters

The objective of these studies was to examine whether or not employment as a fire fighter increases the risk of cancer. Most of the studies are retrospective cohort mortality studies of firefighters and present results for specific cancers as SMRs.

Key Studies

A retrospective cohort mortality study conducted by Baris et al. (2001) was particularly valuable because it quantified exposure as number of runs made by each individual firefighter. Cancer mortality among 7,789 Philadelphia firefighters was compared with the general white male U.S. population. On average, the firefighters worked for 18 years, with 26 years follow-up. There was a significant excess risk of colon cancer (SMR 1.51, 95% CI 1.18–1.93) but not other cancers. After stratification by duration of employment, cumulative number of runs, and number of runs in the first 5 years of employment, there were no positive exposure–response trends for any outcome. Stratification showed significantly increased risks of certain cancers for those employed 9 years or less (all cancers, colon, pancreas, lung, prostate), or more than 20 years (colon, kidney, multiple myeloma); those employed at engine companies (all cancers, buccal cavity and pharynx, colon, multiple myeloma) or ladder companies (leukemia), but not at engine and ladder companies; those first hired before 1935 (buccal cavity and pharynx), between 1935 and 1944 (all cancers, colon, kidney, non-Hodgkin's lymphoma [NHL]) and after 1944 (colon); and those having made fewer than 3,323 runs (colon, NHL), 3,323 to 5,099 runs (colon, skin), but not for more than 5,099 runs.

Aronson et al. (1994) conducted a retrospective cohort mortality study of 5,414 firefighters in metropolitan Toronto. The authors compared the firefighter cohort's mortality rate to that of the Ontario male population. The average time of follow-up was 21 years and the average term of employment was 20 years. There was a significant excess risk of brain and other nervous system cancers (SMR 2.01, 95% CI 1.10–3.37) based on 14 cases, and other malignant neoplasms based on 20 deaths (SMR 2.38, 95% CI 1.45–3.67) but not for other cancer sites. Other malignant neoplasms remained significant for several strata (30 or more years since first employment, less than 15 and more than 30 years of employment, and both age strata [less than 60 years or 60 or more years]) whereas brain and other nervous system cancers were not significantly elevated after stratification.

Bates et al. (2007) analyzed 804,107 male cancer cases from the California Cancer Registry diagnosed from 1988–2003. Firefighters (3,659 cases) were compared to other registrants with adjustments for age, year of diagnosis, ethnicity, and socioeconomic status; the analysis to was restricted to those aged 21–80 years old at diagnosis. Increased risks were reported for testicular cancer (OR 1.54; 95% CI 1.18–2.02), melanoma (OR 1.50; 95% CI 1.33–1.70), brain cancer (OR 1.35, 95% CI 1.06–1.72), prostate cancer (OR 1.22, 95% CI 1.12–1.33), and esophageal cancer (OR 1.48; 95% CI 1.14–1.91). Disease misclassification is an issue because the occupation and industry fields in the registry were self-reported and sometimes were blank, some registrants might have been volunteer firemen, and no employment duration or other way to quantify exposure was available.

Beaumont et al. (1991) conducted a retrospective cohort mortality study of 3,066 firefighters employed in San Francisco, California, between 1940 and 1970. The authors used the U.S. white male population as the reference population. Decreased risk was found for prostate cancer (SMR 0.38, 95% CI 0.16–0.75). There was a significant increased risk for esophageal cancer (SMR 2.04, 95% CI: 1.05–3.57), however, the other cancers were all nonsignificant. There was no adjustment for smoking or alcohol.

Demers et al. (1992b) studied the mortality of 4,546 men employed as firefighters in Seattle and Tacoma, Washington, and Portland, Oregon. Mortality rates in the study population were compared with the national mortality rates for U.S. males. There was no evidence of excess risk of overall mortality from cancer except for a significant increase in cancer mortality from brain tumors (SMR 2.09, 95% CI 1.3–3.2). Younger firefighters (< 40 years of age) appeared to have an excess risk of brain cancer (SMR 3.75, 95% CI 1.2–8.7). The risk of lymphatic and haematopoetic cancers was greatest for men with at least 30 years of employment exposure (SMR 2.05, 95% CI 1.1–3.6), especially for leukemia (SMR 2.60, 95% CI 1.05–4). Demers et al. (1994) conducted a later study of cancer incidence in firefighters that was limited to Seattle and Tacoma, Washington, using tumor registry data.

Duration of active duty was assignable for Seattle firefighters and used as a surrogate measure of cumulative exposure to combustion products from fires; no exposure was assigned for years spent in administrative duties or support services. Total years of employment had to be used for Tacoma firefighters because records identifying the start and end dates of specific duties were not available for all of them. The study population included 2,447 firefighters and was followed from 1974 to 1989. The cancer incidence in this firefighter population was compared with local cancer incidence rates, and with incidence in 1,878 policemen from the same cities. Compared with local rates, firefighters had an elevated rate of prostate cancer (standardized incidence ratio [SIR] 1.5, 95% CI 1.1–1.7), but no significant risks were observed for other cancers. Compared to policemen, no cancers were significantly elevated among firefighters.

Guidotti et al. (1993) conducted a retrospective cohort mortality study of 3,328 firefighters employed between 1927 and 1987 in Edmonton and Calgary, Alberta. An exposure opportunity index term, reflecting estimates of the relative time spent in close proximity to fires by job classification, was used to refine exposure data based on years of service. The authors found a significant increase for all cancers combined (SMR 1.27, 95% CI 1.02–1.55), but no excess risk for any specific cancer.

Using Massachusetts Cancer Registry data from 1986-2003, Kang et al. (2008) reported positive associations for brain cancer (standardized mortality odds ratio [SMOR] 1.90, 95% CI 1.10–3.26) and colon cancer (SMOR 1.36, 95% CI 1.04–1.79) in firefighters compared with police officers, adjusted for age and smoking status. No significant increases in mortality were seen for bladder cancer, kidney cancer, and Hodgkin's lymphoma in firefighters, nor was there a significant association for cancer mortality in firefighters compared with all other occupations. The registry contained 258,964 eligible cancer cases, including 2,125 firefighters. This registry was previously analyzed by Sama et al. (1990), who reported excess age-adjusted ORs for melanoma (SMOR 2.92, 95% CI 1.70–5.03) and bladder cancer (SMOR 1.59, 95% CI 1.02–2.50) in firefighters compared to the rest of the state. Using policemen as the reference group, bladder cancer (SMOR 2.11, 95% CI 1.07–4.14) and NHL (SMOR 3.27, 95% CI 1.19–8.98) were elevated in firefighters but melanoma was not, although it remained significantly elevated among 55–74 year olds. Smoking patterns were compared between firefighters, policemen, and other Massachusetts males, but the SMORs were not adjusted for smoking.

Tornling et al. (1994) conducted a study of cancer incidence and mortality in firefighters in Stockholm, Sweden. An index of the number of fires fought was calculated for each individual. There were no significant increases in cancer-specific mortality, although there was some evidence for associations with brain and stomach cancers. The most highly exposed subgroup (> 1,000 fires fought) had a significant increase in brain cancer mortality (SMR 4.96, CI 1.35–12.70) based on four cases. The excess was not significant for brain cancer incidence in this subgroup (SIR 2.01, CI 0.40–5.88) based on three cases. Stomach cancer incidence was elevated overall (SIR 1.92; 95% CI 1.14–3.04). In this study, the category of high number of runs would be considered low in the Philadelphia firefighters study of Baris et al. (2001); thus, this cohort had relatively low exposure.

Vena and Fiedler (1987) conducted a retrospective cohort study of 1,867 white male firefighters in Buffalo, New York. The firefighter cohort's mortality experience was compared with that of the U.S. population of white men and also presented SMRs by duration of employment. There were elevated SMRs for colon cancer (SMR 1.83, 95% CI 1.05–2.97), and bladder cancer (SMR 2.86, 95%CI 1.30–5.40). In the subgroup of firefighters employed more than 40 years, the SMRs increased to 4.71 and 5.71, both significant, for colon and bladder cancer, respectively.

Supporting Studies

Several other cohort studies on cancer incidence and mortality found no increased risk of cancer among firefighters. Bates et al. (2001) reported no increase in the incidence of any specific cancer site or for all cancers combined among 4,305 firefighters in New Zealand from 1977 to 1995 (1996 for cancer) except for testicular cancer (SIR 2.97, 95% CI 1.3–5.9) between 1990 and 1996, and no increased cancer mortality for any site from 1977 to 1995. After 14 years of follow-up of Parisian firefighters, Deschamps et al. (1995) found nonsignificant excess mortality for a number of outcomes including genitourinary cancer (SMR 3.29), digestive cancer (SMR 1.14), and respiratory cancer (SMR 1.12). Among male firefighters in Seattle employed for at least 1 year between 1945 and 1980, Heyer et al. (1990) found cancer mortality to be similar to the general U.S. white male population.

However, for those over 65 years of age, the SMR was 1.77 (95% CI 1.05–2.79); for those exposed for 30 years or more, the SMR for leukemia was 5.03 (95% CI 1.04–14.7) and 9.89 (95% CI 1.20–35.71) for other lymphatic and hematopoietic cancers. Ma et al. (2006) investigated incident cancer cases among 36,813 firefighters in Florida from 1981–1999 compared with the state's general population. After adjusting for age and calendar year, significantly elevated risks were seen in men for bladder cancer (SIR 1.29, 95% CI 1.01–1.62), thyroid cancer (SIR 1.77, 95% 1.08–2.76), and testicular cancer (SIR 1.6, 95% CI 1.20–2.09). Among women, increases were reported for thyroid cancer (SIR 3.97, 95% CI 1.45–8.65) and Hodgkin's disease (SIR 6.25, 95% CI 1.26–18.3). For male firefighters, significantly decreased risks were found for buccal cancers; digestive cancers, specifically stomach cancer; respiratory cancers, specifically lung cancer; and lymphopoietic cancers.

Using the 27 state National Occupational Mortality Surveillance database from 1984–1990, Burnett et al. (1994) found an elevated PMR for all cancers combined (1.10, 95% CI 1.06–1.14) and for those cases who died before 65 years of age the PMR was 1.12 (95% CI 1.04–1.21)among those ever employed as firefighters. For all CNS cancer deaths, the PMR was 1.03 (95% CI 0.73–1.41) and for those who died before age 65, the PMR was 0.85 (95% CI 0.52–1.34). The strength of this study is the large numbers of cancer deaths in the cohort. A related study, reported by Ma et al. (1998) used a database overlapping that of Burnett et al. (1994) to study racial differences in susceptibility to cancer among firefighters using death certificates from 24 states. An increase of CNS cancer deaths was seen for black firefighters (MOR 6.9, 95% CI 3.0–16.0) but not white firefighters. The overlap between the two studies precludes their consideration as independent investigations of CNS cancer in the white population.

Carozza et al. (2000) conducted and Krishnan et al. (2003) later expanded a population-based case control study on glioma incidence in the San Francisco bay area. The job histories of 476 (Carozza et al. 2000) and 879 (Krishnan et al. 2003) gliomas cases were identified. Both studies found "suggestive (but not significant) evidence" of higher risk of gliomas among firefighters.

The following studies lacked adequate power to identify a relationship between cancer and occupation as a firefighter. Elci et al. (2003) examined job histories of 1,354 male lung cancer cases in a large cancer therapy center in Marmara, Turkey. They found "strong evidence" of increased risk of lung cancer among firefighters, but adjustment for smoking was by smoking status (ever/never) only. Firth et al. (1996) conducted a retrospective study assessing all cancers combined and cancer-specific incidence in males by occupation in a population in New Zealand. They also found "strong evidence" of increased risk for cancer of the larynx among firefighters based on three and four cases of laryngeal cancer observed for firefighters ages 15–54 and 15–64, respectively, but there was no adjustment for smoking.

Several reviews and meta-analyses have assessed cancer risk among firefighters. A meta-analysis conducted by Howe and Burch (1990) concluded there was no increased risk of overall cancer, lung cancer, or colon cancer, but the pooled risk estimates were significantly elevated for brain cancer and multiple myeloma, indicating the possibility of an association. While the pooled risk estimate was significantly elevated for melanoma, the authors did not believe the evidence supported a causal association. To address the accumulating body of literature since the Howe and Burch publication, LeMasters et al. (2006) conducted a meta-analysis of 28 studies of firefighters for selected cancers to determine probable, possible, or unlikely risks. The likelihood of cancer risks for multiple myeloma, NHL, and prostate cancer were determined to be probable while cancers of the testes, skin, malignant melanoma, brain, rectum, buccal cavity and pharynx, stomach, colon and leukemia were classified as possible. Another meta-analysis of 16 studies by Youakim (2006) found significantly increased risks for kidney cancer, NHL mortality, and bladder cancer incidence. Furthermore, among studies having duration of employment information, significantly increased risks were found for mortality from kidney cancer with more than 20 years of employment; brain cancer, colon cancer, and leukemia with more than 30 years of employment; and bladder cancer with more than 40 years of employment as a firefighter.

Three reviews provide very different interpretations of cancer risk among firefighters. In the most recent review, Haas et al. (2003) reviewed 17 studies and found that there was no evidence of an association between employment as a firefighter and cancer mortality; however, the authors noted that consistently elevated rates of brain cancer warrant further investigation. In contrast, Golden et al. (1995) reviewed 19 studies and concluded that firefighters are at increased risk of leukemia, NHL, multiple myeloma, brain cancer, and bladder cancer with weak evidence supporting risks of rectal cancer, colon cancer, stomach cancer, prostate cancer, and melanoma.

On the basis of 21 studies, Guidotti (1995) concluded there is sufficient evidence for a presumption of increased risk among firefighters for lung cancer, genitourinary cancers, and colon and rectal cancers. Associations for brain cancer, and lymphatic and hematopoietic cancers are less clear.

In 2010, the International Agency for Research on Cancer (IARC) released a monograph including an assessment of the carcinogenicity of firefighting as an occupation. After a review of 42 epidemiologic studies and meta-analyses, IARC found increased risks for testicular cancer, prostate cancer, and NHL. The assessment concluded that "occupational exposure as a firefighter is *possibly carcinogenic to humans (Group 2B)*," based on "*limited evidence* in humans" (IARC 2010).

Cancer in Incinerator Workers and Surrounding Communities

The committee reviewed several community-based studies of cancer incidence in populations living near municipal solid waste incinerators. In these studies, exposure is measured as the distance between the place of residence and the incinerator, or as area-level dioxin concentrations at ground level estimated from air dispersion models. In the community studies, the focus is primarily on whether cancer rates are elevated in populations that live within a specified distance (radius) of an incinerator.

Key Studies

No key studies describing cancer associated with incinerators were identified by the committee.

Supporting Studies

Increasing distance from an incinerator is generally associated with a decreasing risk of cancer; however, cancer risks are not necessarily elevated nearest the incinerator, as demonstrated by the following studies. Elliott et al. (1996) conducted a retrospective study of cancer incidence in a population of over 14 million people living near 72 solid waste incinerators in Great Britain to investigate whether cancer incidence was associated with exposure to combustion products from the incinerators. There was a significant ($p < 0.05$) decline in cancer incidence associated with increasing distance from the incinerators for all cancers, as well as stomach, colorectal, liver, and lung cancer after adjustment for age, sex, and geographic region. No associations were seen for nasal and nasopharyngeal, larynx, connective tissue, or lymphatic and hematopoietic cancers. In a population-based case-control study of lung cancer, Biggeri et al. (1996) measured distance from four sources of exposure—a shipyard, iron foundry, incinerator, and the center of the city of Trieste, Italy. There were 755 histologically confirmed cases of lung cancer in men. Controls were matched by age and month of death. After adjusting for age, smoking habit, occupational exposure, and levels of PM, the excess relative risk (6.7 at the incinerator) declined with increasing distance. In a small community-based mortality study in Rome, Michelozzi et al. (1998) found no association between proximity to a waste incinerator and all cancers combined or most specific cancers after adjusting for age, gender, and socioeconomic status.

In a French ecologic study, a significant association was found between living near an incinerator, and soft-tissue sarcoma (STS) and NHL (Viel et al. 2000). Highly significant clusters for both STS and NHL ($p < 0.05$) were observed in the area around the municipal solid waste incinerator. A population-based case-control study confirmed an increased NHL risk (OR 2.3, 95% CI 1.4–3.8) for individuals living in the area with the highest dioxin concentration compared with those in the area with the lowest concentration (Floret et al. 2003). A follow-up study on a larger geographical area in France, found evidence of an association between dioxin exposure from a waste incinerator and increased risk of NHL ($p = 0.04$) after controlling for geographic area, industrial pollution, years of polluting industry, and population density (Viel et al. 2008).

One review examined reproductive effects, respiratory effects, and cancer in residential areas around incinerators and mortality and biomarkers among incinerator workers (Hu and Shy 2001). The authors report inconclusive findings for cancer. Generally these occupational studies are hampered by low statistical power, and the residential studies must account for multiple sources of exposure to potential carcinogens.

The committee also reviewed studies on long-term health outcomes for incinerator workers. Gustavsson et al.

(1993) conducted a retrospective cohort study of 5,542 chimney sweeps, 176 incinerator workers, 296 gas workers, and 695 bus garage workers. In the small cohort of incinerator workers, the SMR for esophageal cancer was 1.50, based on one case. Comba et al. (2003) conducted a case-control study in Mantua, Italy, and found a significant increase in risk of STS associated with living within 2 km of an industrial waste incinerator, after controlling for age and gender. However, when stratified by residential distance from the incinerator, five cases lived within 2 km of the incinerator (OR 31.4, 95% 5.6–176.1), but no other distances showed an increase in STS risk. A larger case-control study, by Zambon et al. (2007) of STS and residential history in a large area with 10 municipal solid waste incinerators in Veneto, Italy, found a significant increase in the risk of sarcoma on the basis of both the level and the length of modeled exposure to dioxin-like substances. Among the 172 cases and 405 controls, those with the longest and highest exposure had the highest sarcoma risk (OR 3.30, 95% CI 1.24–8.76). No adjustments were made for socioeconomic status or other potential confounders.

Cancer in Gulf War Veterans Exposed to Oil-Well–Fire Smoke

The IOM has previously evaluated the scientific literature on associations between illness and exposure to toxic agents associated with Gulf War service. Of particular interest to the committee is the IOM's previous review of Gulf War studies of veterans exposed to smoke from oil-well fires. The current committee did not review the studies below; rather, it relied on the assessments of the earlier IOM committees who have examined the evidence and weighed the health effects literature, including cancer, related to this exposure (IOM 2005, 2006, 2010). Two assessments of brain cancer and one of respiratory cancer were considered relevant to this report.

Bullman et al. (2005) explored the relationship between estimated exposure to sarin gas during chemical munitions destruction at Khamisiyah, Iraq, in 1991 with cause-specific mortality of Gulf War veterans through December 31, 2000. Using the DoD's 2000 sarin plume exposure model, 100,487 military personnel were identified as potentially exposed and 224,980 similarly deployed military personnel were considered unexposed. The study found an increased risk of brain cancer deaths in the exposed population (RR 1.94, 95% CI 1.12–3.34; 25 exposed cases vs. 27 unexposed cases), and there was a suggestion of a dose-response relationship with increased risk among those who were considered exposed for 2 days (6 cases) relative to 1 day (19 cases) (RR 3.26, 95% CI 1.33–7.96 and RR 1.72, 95% CI 0.95–3.10, respectively). The authors also discussed modeling exposure to smoke from oil-well fires as a confounder, and the effect estimates for exposure to Khamisiyah nerve agents remained elevated. There was no significant elevation in risk associated with exposure to oil-well fires as a main effect. Because brain cancer likely has a latent period of 10–20 years and Bullman et al. (2005) had fewer than 9 years of follow-up, it was concluded that additional follow-up is needed to draw any definitive conclusions concerning the association between exposure to oil-well fire smoke and the development of brain cancer.

Studying a broader cohort of military personnel that included the brain cancer cases discussed by Bullman et al. (2005), Barth et al. (2009) identified 144 brain cancer deaths among 621,902 Gulf War deployed veterans and 228 brain cancer deaths among 746,248 nondeployed veterans followed through 2004. There were 123,478 veterans exposed to oil-well–fire smoke based on a definition of exposure to 0.26 mg/m^3 or more of total suspended particulates for at least 1 day. Smoke exposure and brain cancer risk was modeled with and without modeled exposure to nerve agents (both as exposed vs. not exposed) while controlling for age, sex, race, and unit type. Without nerve agent exposure, the relative risk for brain cancer associated with oil-well–fire smoke exposure was 1.67 (95% CI 1.05–2.65) and 1.81 (95% CI 1.00–3.27) when modeled with 2 or more days of nerve agent exposure (IOM 2010).

The *Gulf War and Health* report on combustion products (IOM 2005) discussed one study (Smith et al. 2002) that assessed respiratory cancer in Gulf War veterans exposed to smoke from oil-well fires.[4] The report stated:

> The study population consisted of 405,142 regular active-duty U.S. military personnel who were in the gulf region during the Kuwaiti oil-well fires. Hospitalization records were examined from DoD military treatment facilities from August 1, 1991, until hospitalization, separation from active-duty service, or July 31, 1999. Modeling was used to estimate troop exposure to oil-well fire smoke. The risk of malignant neoplasms of the respiratory and intrathoracic organs was modestly increased in the exposed group, but the CI included the null (adjusted RR 1.10, 95% CI 0.56–2.17). The relatively short observation period (8 years) is a limitation of this study for assessing cancer risk.

---

[4]The following text was excerpted from IOM (2005).

Cancer in OEF/OIF Veterans Exposed to Burn Pits

No epidemiologic investigations of cancer outcomes among OEF/OIF veterans specifically exposed to burn pits were available to the committee. However, accounts of cancer presumed to be attributable to burn pit emission have been described by the popular press (Risen 2010).

Conclusions

A few cancer sites in particular came to the committee's attention. Brain, colon, and testicular cancer were reported at increased risks in some studies. The committee carefully considered the evidence for an association between combustion products and those cancers in firefighters, incinerator workers, and Gulf War soldiers.

There were two key cohort studies that reported positive and significant results for brain cancer and occupational exposure. In a large study of firefighters in four cities in Washington state, Demers et al. (1992b) found an excess risk of brain cancer (SMR 2.09, 95% CI 1.3–3.2). The SMR was higher (3.75, (95% CI 1.2–8.8) among firefighters less than 40 years of age, but there were no trends with duration of employment, and results were not corroborated later by Demers et al. (1994). Tornling et al. (1994) found an SMR for brain cancer of 4.96 (95% CI 1.35–12.70) based on four cases in a cohort mortality study of Swedish firefighters who had fought more than 1,000 fires and after more than 30 years of latency, but for the total cohort the SMR was 2.79 (95% CI 0.91–6.51). There were also two key registry-based case-control studies that found significant positive associations for brain cancer. Bates et al. (2007) reported an OR of 1.35 (95% CI 1.06–1.72) for firefighters relative to other occupations, based on an alternative control group selected from cases of all other cancers excluding several specific sites. Kang et al. (2008) found an SMOR of 1.90 (95% CI 1.1–2.36) using policemen as the reference group. The largest cohort study of firefighters (Baris et al. 2001), however, found no excess risk of brain cancer in firefighters. The overall SMR of 0.61 for brain cancer was based on eight cases, and there were no trends with increasing number of runs or duration of employment. In summary, there are two cohort studies (Demers et al. 1992b; Tornling et al. 1994) and two registry-based case-control studies (Bates et al. 2007; Kang et al. 2008) with elevated risk estimates for brain cancer in firefighters, all with several limitations. The largest cohort study, and the only one with a quantitative exposure assessment based on number of runs, is negative.

Evidence for colon cancer is mixed. The strongest key study (Baris et al. 2001) reported a significantly increased SMR for colon cancer of 1.51(95% CI 1.18–1.93) as did Elliott et al. (1996) (SMR = 1.04, 95% CI 1.02–1.06), Kang et al. (2008) (SMOR = 1.36, 95% CI 1.04–1.79), and Vena and Fiedler (1987) (SMR = 1.83, 95% CI 1.05–2.97). Elevated but not significant risks were reported by Demers et al. (1994) (IDR = 1.3, 95% CI 0.6–1.3) and Giudotti (1993) (SMR = 1.61, 95% CI 0.88–2.71). Several other key studies, however, found no evidence of excess risk.

Several studies reported significant risks of testicular cancer among firefighters. Three of the four key studies were positive: Bates et al. (2007) found an OR of 1.54 (95% CI 1.18–2.02), Kang et al. (2008) reported an SMOR of 1.53 (95% CI 0.75–3.14), and Aronson reported an SMR of 2.52 (95% CI 0.52–7.37), but Beaumont et al. (1991) found an SMR of 0.40 (95% CI 0.18–0.77). One supporting study by Ma et al. (2006) has an SIR for testicular cancer of 1.6 (95% CI 1.20–2.09). Similarly inconsistent significant results were reported for prostate cancer. Three key studies of disease risks among firemen reported significantly different risks than the selected control populations. Bates et al. (2007) reported an OR of 1.22 (95% CI 1.12–1.33); Beaumont et al. (1991) reported a SMR of 0.38 (95% CI 0.16–0.75); and Demers et al. (1994) reported a SIR of 1.4 (95% CI 1.1–1.7)

In summary, a review of the firefighter literature on specific cancer risks is generally inconsistent. Although there is some evidence of a positive association between firefighting and cancer of the brain, testes, and colon, it does not rise to the level of limited/suggestive for any particular cancer in light of the major study design constraints noted above. Even the best studies are limited by inadequate exposure assessment and the inappropriate selection of the general population as the reference group. Though several key studies included internal exposure-response analyses, none of these results were positive for any cancer.

The studies of firefighters, incinerator workers, and communities near incinerators do not show consistent increased risks for any specific cancer. While a few studies reported increased risks of brain, testicular, and prostate cancer, there is a lack of evidence for exposure-response relationships (based on duration of employment or number of fires fought); thus, the studies cannot be directly linked to exposure to combustion products. In the absence of

TABLE 6-2 Risk Estimates by Cancer Site[a]

Study	Study Type	Population	Risk Estimate (95% CI)	
All cancers				
Aronson et al. 1994	K	Firefighters	SMR	1.05 (0.91–1.20)
Baris et al. 2001	K	Firefighters	SMR	*1.10 (1.00–1.20)*
Beaumont et al. 1991	K	Firefighters	SMR	0.95 (0.84–1.08)
Demers et al. 1994	K	Firefighters	SIR	1.1 (0.9–1.2)
Demers et al. 1994	K	Firefighters	IDR	1.0 (0.8–1.3)
Guidotti 1993	K	Firefighters	SMR	*1.27 (1.02–1.55)*
Tornling et al. 1994	K	Firefighters	SMR	1.02 (0.88–1.25)
Vena and Fiedler 1987	K	Firefighters	SMR	1.09 (0.89–1.32)
Bates et al. 2001	S	Firefighters	SIR	0.95 (0.8–1.1)
Burnett et al. 1994	S	Firefighters	PMR	*1.10 (1.06–1.14)*
Deschamps et al. 1995	S	Firefighters	SMR	0.89 (0.53–1.40)
Heyer et al. 1990	S	Firefighters	SMR	0.96 (0.77–1.18)
Ma et al. 2006	S	Firefighters	SIR	*0.84 (0.79–0.90)*
Elliott et al. 1996	S	Incinerator communities	SMR	*1.04 (1.03–1.04)*
Michelozzi et al. 1998	S	Incinerator communities	SMR	0.88 (0.60–1.26)
Gustavsson et al. 1989	S	Incinerator workers	SMR	1.07 (0.67–1.62)
Bullman et al. 2005		Gulf War	RR	0.97 (0.82–1.16)
Smith et al. 2002		Gulf War	RR	*0.93 (p > 0.05)*
Oral and Pharynx				
Baris et al. 2001	K	Firefighters	SMR	1.36 (0.87–2.14)
Beaumont et al. 1991	K	Firefighters	SMR	1.43 (0.71–2.57)
Demers et al. 1994	K	Firefighters	SIR	1.1 (0.6–2.0)
Demers et al. 1994	K	Firefighters	IDR	0.8 (0.3–1.9)
Guidotti 1993	K	Firefighters	SMR	1.14 (0.14–4.10)
Kang et al. 2008	K	Firefighters	SMOR	0.72 (0.37–1.41)
Ma et al. 2006	S	Firefighters	SIR	*0.67 (0.47–0.91)*
Lip				
Beaumont et al. 1991	K	Firefighters	SMR	6.17 (0.75–22.29)
Kang et al. 2008	K	Firefighters	SMOR	1.10 (0.24–5.06)
Tongue				
Beaumont et al. 1991	K	Firefighters	SMR	1.06 (0.13–3.86)
Pharynx				
Aronson et al. 1994	K	Firefighters	SMR	1.39 (0.38–3.57)
Beaumont et al. 1991	K	Firefighters	SMR	1.17 (0.32–3.00)
Kang et al. 2008	K	Firefighters	SMOR	1.17 (0.19–7.17)
Deschamps et al. 1995	S	Firefighters	SMR	0.81 (0.10–2.93)
Digestive				
Beaumont et al. 1991	K	Firefighters	SMR	*1.27 (1.04–1.55)*
Vena and Fiedler 1987	K	Firefighters	SMR	1.38 (0.98–1.89)
Deschamps et al. 1995	S	Firefighters	SMR	1.14 (0.37–2.66)
Heyer et al. 1990	S	Firefighters	SMR	1.06 (0.71–1.52)
Ma et al. 2006	S	Firefighters	SIR	*0.76 (0.65–0.89)*
Esophagus				
Aronson et al. 1994	K	Firefighters	SMR	0.40 (0.05–1.43)
Baris et al. 2001	K	Firefighters	SMR	0.56 (0.25–1.24)
Bates et al. 2007	K	Firefighters	OR	*1.48 (1.14–1.91)*
Beaumont et al. 1991	K	Firefighters	SMR	*2.04 (1.05–3.57)*
Demers et al. 1994	K	Firefighters	SIR	1.3 (0.4–3.3)
Kang et al. 2008	K	Firefighters	SMOR	0.93 (0.61–1.41)
Vena and Fiedler 1987	K	Firefighters	SMR	1.34 (0.27–3.91)
Bates et al. 2001	S	Firefighters	SIR	1.67 (0.3–4.9)

continued

TABLE 6-2 Continued

Study	Study Type	Population	Risk Estimate (95% CI)	
Heyer et al. 1990	S	Firefighters	SMR	0.44 (0.01–2.50)
Ma et al. 2006	S	Firefighters	SIR	0.62 (0.31–1.11)
Gustavsson et al. 1993	S	Incinerator workers	SMR	1.50 (0.04–8.34)
Stomach				
Aronson et al. 1994	K	Firefighters	SMR	0.51 (0.20–1.05)
Baris et al. 2001	K	Firefighters	SMR	0.90 (0.61–1.35)
Bates et al. 2007	K	Firefighters	OR	0.80 (0.61–1.07)
Beaumont et al. 1991	K	Firefighters	SMR	1.31 (0.82–1.99)
Demers et al. 1994	K	Firefighters	SIR	1.4 (0.6–2.7)
Demers et al. 1994	K	Firefighters	IDR	0.4 (0.1–1.2)
Guidotti 1993	K	Firefighters	SMR	0.81 (0.30–1.76)
Kang et al. 2008	K	Firefighters	SMOR	0.83 (0.53–1.29)
Tornling et al. 1994	K	Firefighters	SMR	1.21 (0.62–2.11)
Vena and Fiedler 1987	K	Firefighters	SMR	1.19 (0.48–2.46)
Bates et al. 2001	S	Firefighters	SIR	0.76 (0.2–2.2)
Heyer et al. 1990	S	Firefighters	SMR	1.13 (0.41–2.47)
Ma et al. 2006	S	Firefighters	SIR	***0.5 (0.25–0.90)***
Elliott et al. 1996	S	Incinerator communities	SMR	***1.05 (1.03–1.08)***
Gustavsson et al. 1989	S	Incinerator workers	SMR	1.32 (0.27–3.86)
Intestine				
Beaumont et al. 1991	K	Firefighters	SMR	0.99 (0.63–1.47)
Heyer et al. 1990	S	Firefighters	SMR	0.79 (0.32–1.64)
Colon				
Aronson et al. 1994	K	Firefighters	SMR	0.60 (0.30–1.08)
Baris et al. 2001	K	Firefighters	SMR	***1.51 (1.18–1.93)***
Bates et al. 2007	K	Firefighters	OR	0.90 (0.79–1.03)
Demers et al. 1994	K	Firefighters	SIR	1.1 (0.7–1.6)
Demers et al. 1994	K	Firefighters	IDR	1.3 (0.6–3.0)
Guidotti 1993	K	Firefighters	SMR	1.61 (0.88–2.71)
Kang et al. 2008	K	Firefighters	SMOR	***1.36 (1.04–1.79)***
Tornling et al. 1994	K	Firefighters	SMR	0.85 (0.31–1.85)
Vena and Fiedler 1987	K	Firefighters	SMR	***1.83 (1.05–2.97)***
Bates et al. 2001	S	Firefighters	SIR	0.60 (0.2–1.2)
Ma et al. 2006	S	Firefighters	SIR	1.16 (0.92–1.45)
Elliott et al. 1996	S	Incinerator communities	SMR	***1.04 (1.02–1.06)***
Bullman et al. 2005		Gulf War	RR	1.17 (0.61–2.25)
Cecum				
Bates et al. 2007	K	Firefighters	OR	1.09 (0.82–1.44)
Rectum				
Aronson et al. 1994	K	Firefighters	SMR	1.71 (0.91–2.93)
Baris et al. 2001	K	Firefighters	SMR	0.99 (0.59–1.68)
Beaumont et al. 1991	K	Firefighters	SMR	1.45 (0.77–2.49)
Demers et al. 1994	K	Firefighters	SIR	1.0 (0.5–1.8)
Demers et al. 1994	K	Firefighters	IDR	1.3 (0.5–3.9)
Kang et al. 2008	K	Firefighters	SMOR	0.86 (0.58–1.26)
Tornling et al. 1994	K	Firefighters	SMR	2.07 (0.89–4.08)
Vena and Fiedler 1987	K	Firefighters	SMR	2.08 (0.83–4.28)
Bates et al. 2001	S	Firefighters	SIR	1.15 (0.5–2.2)
Burnett et al. 1994	S	Firefighters	PMR	***1.48 (1.05–2.05)***
Heyer et al. 1990	S	Firefighters	SMR	0.65 (0.08–2.37)
Ma et al. 2006	S	Firefighters	SIR	0.88 (0.56–1.32)

TABLE 6-2 Continued

Study	Study Type	Population	Risk Estimate (95% CI)	
Gustavsson et al. 1989	S	Incinerator workers	SMR	2.32 (0.28–8.37)
Liver				
Aronson et al. 1994	K	Firefighters	SMR	0.84 (0.10–3.05)
Baris et al. 2001	K	Firefighters	SMR	0.82 (0.41–1.64)
Beaumont et al. 1991	K	Firefighters	SMR	1.91 (0.87–3.63)
Kang et al. 2008	K	Firefighters	SMOR	1.15 (0.55–2.41)
Tornling et al. 1994	K	Firefighters	SMR	1.49 (0.41–3.81)
Vena and Fiedler 1987	K	Firefighters	SMR	0.98 (0.11–3.52)
Ma et al. 2006	S	Firefighters	SIR	0.74 (0.32–1.46)
Elliott et al. 1996	S	Incinerator communities	SMR	***1.13 (1.05–1.22)***
Pancreas				
Aronson et al. 1994	K	Firefighters	SMR	1.40 (0.77–2.35)
Baris et al. 2001	K	Firefighters	SMR	0.96 (0.64–1.44)
Bates et al. 2007	K	Firefighters	OR	0.90 (0.70–1.17)
Beaumont et al. 1991	K	Firefighters	SMR	1.25 (0.73–2.00)
Demers et al. 1994	K	Firefighters	SIR	1.1 (0.4–2.3)
Demers et al. 1994	K	Firefighters	IDR	1.1 (0.3–5.5)
Guidotti 1993	K	Firefighters	SMR	1.55 (0.50–3.62)
Kang et al. 2008	K	Firefighters	SMOR	0.86 (0.53–1.40)
Tornling et al. 1994	K	Firefighters	SMR	0.84 (0.27–1.96)
Vena and Fiedler 1987	K	Firefighters	SMR	0.38 (0.04–1.36)
Bates et al. 2001	S	Firefighters	SIR	1.28 (0.3–3.7)
Ma et al. 2006	S	Firefighters	SIR	0.57 (0.30–1.10)
Bullman et al. 2005		Gulf War	RR	0.82 (0.39–1.73)
Respiratory				
Beaumont et al. 1991	K	Firefighters	SMR	0.83 (0.64–1.06)
Vena and Fiedler 1987	K	Firefighters	SMR	0.94 (0.62–1.36)
Deschamps et al. 1995	S	Firefighters	SMR	1.12 (0.45–2.30)
Heyer et al. 1990	S	Firefighters	SMR	1.01 (0.65–1.34)
Ma et al. 2006	S	Firefighters	SIR	0.67 (0.57–0.78)
Sinus				
Demers et al. 1994	K	Firefighters	SIR	2.2 (0.1–12.4)
Larynx				
Aronson et al. 1994	K	Firefighters	SMR	0.37 (0.01–2.06)
Baris et al. 2001	K	Firefighters	SMR	0.75 (0.31–1.81)
Beaumont et al. 1991	K	Firefighters	SMR	0.80 (0.17–2.35)
Demers et al. 1994	K	Firefighters	SIR	1.0 (0.3–2.3)
Demers et al. 1994	K	Firefighters	IDR	0.8 (0.2–3.5)
Kang et al. 2008	K	Firefighters	SMOR	0.66 (0.39–1.10)
Firth et al. 1996	S	Firefighters	SIR	***10.74 (2.79–27.76)***
Ma et al. 2006	S	Firefighters	SIR	0.73 (0.44–1.12)
Michelozzi et al. 1998	S	Incinerator communities	SMR	2.36 (0.27–8.00)
Lung				
Aronson et al. 1994	K	Firefighters	SMR	0.95 (0.71–1.24)
Baris et al. 2001	K	Firefighters	SMR	1.13 (0.97–1.32)
Bates et al. 2007	K	Firefighters	OR	0.98 (0.88–1.09)
Beaumont et al. 1991	K	Firefighters	SMR	0.84 (0.64–1.08)
Demers et al. 1994	K	Firefighters	IDR	1.1 (0.6–1.9)
Demers et al. 1994	K	Firefighters	SIR	1.0 (0.7–1.3)
Guidotti 1993	K	Firefighters	SMR	1.42 (0.91–2.11)
Kang et al. 2008	K	Firefighters	SMOR	1.02 (0.79–1.31)

continued

TABLE 6-2 Continued

Study	Study Type	Population	Risk Estimate (95% CI)	
Tornling et al. 1994	K	Firefighters	SMR	0.90 (0.53–1.42)
Bates et al. 2001	S	Firefighters	SIR	1.14 (0.7–1.8)
Burnett et al. 1994	S	Firefighters	PMR	1.02 (0.94–1.11)
Elci et al. 2003	S	Firefighters	OR	*6.8 (1.3–37.4)*
Firth et al. 1996	S	Firefighters	SIR	1.65 (0.60–3.62)
Heyer et al. 1990	S	Firefighters	SMR	0.97 (0.65–1.39)
Ma et al. 2006	S	Firefighters	SIR	*0.65 (0.54–0.78)*
Biggeri et al. 1996	S	Incinerator communities	Excess risk	1.48 (p = 0.0937)
Elliott et al. 1996	S	Incinerator communities	SMR	*1.08 (1.07–1.09)*
Michelozzi et al. 1998	S	Incinerator communities	SMR	0.95 (0.48–1.69)
Gustavsson et al. 1989	S	Incinerator workers	SMR	1.97 (0.90–3.74)
Bullman et al. 2005		Gulf War	RR	0.72 (0.47–1.10)
Bone				
Ma et al. 2006	S	Firefighters	SIR	1.02 (0.27–2.61)
Soft Tissue Sarcoma				
Kang et al. 2008	K	Firefighters	SMOR	1.05 (0.46–2.37)
Ma et al. 2006	S	Firefighters	SIR	1 (0.55–1.69)
Elliott et al. 1996	S	Incinerator communities	SMR	1.03 (0.94–1.13)
Viel et al. 2000	S	Incinerator communities	SIR	*1.44 (p = 0.004)*
Comba et al. 2003	S	Incinerator workers	OR	*31.4 (5.6–176.1)*
Zambon et al. 2007	S	Incinerator workers	OR	*3.30 (1.24–8.76)*
Skin				
Aronson et al. 1994	K	Firefighters	SMR	0.73 (0.09–2.63)
Baris et al. 2001	K	Firefighters	SMR	1.18 (0.64–2.20)
Bates et al. 2007	K	Firefighters	OR	*1.50 (1.33–1.70)*
Beaumont et al. 1991	K	Firefighters	SMR	1.69 (0.68–3.49)
Demers et al. 1994	K	Firefighters	IDR	1.0 (0.4–1.8)
Demers et al. 1994	K	Firefighters	SIR	1.2 (0.6–2.3)
Guidotti 1993	K	Firefighters	SMR	0.0 (0.0–3.31)
Kang et al. 2008	K	Firefighters	SMOR	0.65 (0.44–0.97)
Bates et al. 2001	S	Firefighters	SIR	1.26 (0.8–1.9)
Burnett et al. 1994	S	Firefighters	PMR	*1.63 (1.15–2.23)*
Ma et al. 2006	S	Firefighters	SIR	1.17 (0.95–1.42)
Breast				
Demers et al. 1994	K	Firefighters	SIR	2.4 (0.1–13.3)
Kang et al. 2008	K	Firefighters	SMOR	0.25 (0.03–2.31)
Ma et al. 2006	S	Firefighters	SIR	0.51 (0.06–1.84)
Genito–urinary				
Deschamps et al. 1995	S	Firefighters	SMR	3.29 (0.40–11.88)
Prostate				
Aronson et al. 1994	K	Firefighters	SMR	1.32 (0.76–2.15)
Baris et al. 2001	K	Firefighters	SMR	0.96 (0.68–1.37)
Bates et al. 2007	K	Firefighters	OR	*1.22 (1.12–1.33)*
Beaumont et al. 1991	K	Firefighters	SMR	*0.38 (0.16–0.75)*
Demers et al. 1994	K	Firefighters	SIR	*1.4 (1.1–1.7)*
Demers et al. 1994	K	Firefighters	IDR	1.1 (0.7–1.8)
Guidotti 1993	K	Firefighters	SMR	1.46 (0.63–2.88)
Kang et al. 2008	K	Firefighters	SMOR	0.98 (0.78–1.23)
Tornling et al. 1994	K	Firefighters	SMR	1.21 (0.66–2.02)
Vena and Fiedler 1987	K	Firefighters	SMR	0.71 (0.23–1.65)
Bates et al. 2001	S	Firefighters	SIR	1.08 (0.5–1.9)

TABLE 6-2 Continued

Study	Study Type	Population	Risk Estimate (95% CI)	
Ma et al. 2006	S	Firefighters	SIR	1.1 (0.95–1.42)
Testes				
Aronson et al. 1994	K	Firefighters	SMR	2.52 (0.52–7.37)
Bates et al. 2007	K	Firefighters	OR	*1.54 (1.18–2.02)*
Beaumont et al. 1991	K	Firefighters	SMR	*0.40 (0.18–0.77)*
Kang et al. 2008	K	Firefighters	SMOR	1.53 (0.75–3.14)
Bates et al. 2001	S	Firefighters	SIR	1.55 (0.8–2.8)
Ma et al. 2006	S	Firefighters	SIR	*1.6 (1.20–2.09)*
Bladder				
Aronson et al. 1994	K	Firefighters	SMR	1.28 (0.51–2.63)
Baris et al. 2001	K	Firefighters	SMR	1.25 (0.77–2.00)
Bates et al. 2007	K	Firefighters	OR	*0.85 (0.72–1.00)*
Beaumont et al. 1991	K	Firefighters	SMR	0.57 (0.19–1.35)
Beaumont et al. 1991	K	Firefighters	SMR	0.61 (0.28–1.17)
Demers et al. 1994	K	Firefighters	SIR	1.2 (0.7–1.9)
Demers et al. 1994	K	Firefighters	IDR	1.7 (0.7–4.3)
Guidotti 1993	K	Firefighters	SMR	3.16 (0.86–8.08)
Kang et al. 2008	K	Firefighters	SMOR	1.22 (0.89–1.69)
Vena and Fiedler 1987	K	Firefighters	SMR	*2.86 (1.30–5.40)*
Bates et al. 2001	S	Firefighters	SIR	1.14 (0.4–2.7)
Burnett et al. 1994	S	Firefighters	PMR	0.99 (0.70–1.37)
Ma et al. 2006	S	Firefighters	SIR	*1.29 (1.01–1.62)*
Elliott et al. 1996	S	Incinerator communities	SMR	1.01 (0.98–1.04)
Gustavsson et al. 1989	S	Incinerator workers	SMR	1.39 (0.03–7.77)
Kidney				
Aronson et al. 1994	K	Firefighters	SMR	0.43 (0.05–1.56)
Baris et al. 2001	K	Firefighters	SMR	1.07 (0.61–1.88)
Bates et al. 2007	K	Firefighters	OR	1.07 (0.87–1.31)
Beaumont et al. 1991	K	Firefighters	SMR	0.68 (0.19–1.74)
Demers et al. 1994	K	Firefighters	SIR	0.5 (0.1–1.6)
Demers et al. 1994	K	Firefighters	IDR	0.4 (0.1–2.1)
Guidotti 1993	K	Firefighters	SMR	*4.14 (1.66–8.53)*
Kang et al. 2008	K	Firefighters	SMOR	1.34 (0.90–2.01)
Tornling et al. 1994	K	Firefighters	SMR	1.10 (0.30–2.81)
Vena and Fiedler 1987	K	Firefighters	SMR	1.30 (0.26–3.80)
Bates et al. 2001	S	Firefighters	SIR	0.57 (0.1–2.1)
Burnett et al. 1994	S	Firefighters	PMR	*1.44 (1.08–1.89)*
Ma et al. 2006	S	Firefighters	SIR	0.78 (0.52–1.14)
Michelozzi et al. 1998	S	Incinerator communities	SMR	2.76 (0.31–9.34)
Brain and Nervous System				
Aronson et al. 1994	K	Firefighters	SMR	*2.01 (1.10–3.37)*
Beaumont et al. 1991	K	Firefighters	SMR	0.81 (0.26–1.90)
Vena and Fiedler 1987	K	Firefighters	SMR	2.36 (0.86–5.13)
Burnett et al. 1994	S	Firefighters	PMR	1.03 (0.73–1.41)
Heyer et al. 1990	S	Firefighters	SMR	0.95 (0.26–7.89)
Ma et al. 2006	S	Firefighters	SIR	*0.58 (0.31–0.97)*
Eye				
Demers et al. 1994	K	Firefighters	SIR	5.2 (0.6–18.8)
Ma et al. 2006	S	Firefighters	SIR	1.54 (0.42–3.95)

continued

TABLE 6-2 Continued

Study	Study Type	Population	Risk Estimate (95% CI)	
Brain				
Baris et al. 2001	K	Firefighters	SMR	0.61 (0.31–1.22)
Bates et al. 2007	K	Firefighters	OR	*1.35 (1.06–1.72)*
Demers et al. 1994	K	Firefighters	SIR	1.1 (0.3–2.9)
Demers et al. 1994	K	Firefighters	IDR	1.4 (0.2–11)
Guidotti 1993	K	Firefighters	SMR	1.47 (0.30–4.29)
Kang et al. 2008	K	Firefighters	SMOR	*1.90 (1.10–3.26)*
Tornling et al. 1994	K	Firefighters	SMR	2.79 (0.91–6.51)
Bates et al. 2001	S	Firefighters	SIR	1.27 (0.4–3.0)
Gustavsson et al. 1989	S	Incinerator workers	SMR	2.44 (0.06–13.59)
Barth et al. 2009		Gulf War	RR	0.90 (0.73–1.11)
Bullman et al. 2005		Gulf War	RR	*1.94 (1.12–3.34)*
Glioma				
Carrozza et al. 2000	S	Firefighters	OR	2.7 (0.3–26.1)
Krishnan et al. 2003	S	Firefighters	OR	5.88 (0.70–49.01)
Thyroid				
Bates et al. 2007	K	Firefighters	OR	1.17 (0.82–1.67)
Demers et al. 1994	K	Firefighters	SIR	0.8 (0.2–4.2)
Kang et al. 2008	K	Firefighters	SMOR	0.71 (0.30–1.70)
Ma et al. 2006	S	Firefighters	SIR	*1.77 (1.08–2.73)*
Lymphatic and Hematopoietic				
Aronson et al. 1994	K	Firefighters	SMR	0.98 (0.58–1.56)
Beaumont et al. 1991	K	Firefighters	SMR	0.65 (0.35–1.09)
Beaumont et al. 1991	K	Firefighters	SMR	0.89 (0.24–2.29)
Guidotti 1993	K	Firefighters	SMR	1.27 (0.61–2.33)
Tornling et al. 1994	K	Firefighters	SMR	0.44 (0.09–1.27)
Vena and Fiedler 1987	K	Firefighters	SMR	0.55 (0.18–1.29)
Burnett et al. 1994	S	Firefighters	PMR	*1.30 (1.11–1.51)*
Heyer et al. 1990	S	Firefighters	SMR	1.26 (0.65–2.22)
Ma et al. 2006	S	Firefighters	SIR	*0.68 (0.54–0.85)*
Elliott et al. 1996	S	Incinerator communities	SMR	1.01 (0.99–1.03)
Michelozzi et al. 1998	S	Incinerator communities	SMR	1.20 (0.24–3.37)
Gustavsson et al. 1989	S	Incinerator workers	SMR	1.30 (0.16–4.71)
Hodgkin's Disease				
Aronson et al. 1994	K	Firefighters	SMR	2.04 (0.42–5.96)
Demers et al. 1994	K	Firefighters	SIR	0.7 (0.0–4.1)
Kang et al. 2008	K	Firefighters	SMOR	1.81 (0.72–4.53)
Ma et al. 2006	S	Firefighters	SIR	0.77 (0.38–1.38)
Non-Hodgkin's Lymphoma				
Baris et al. 2001	K	Firefighters	SMR	1.41 (0.91–2.19)
Bates et al. 2007	K	Firefighters	OR	1.07 (0.90–1.26)
Demers et al. 1994	K	Firefighters	IDR	1.8 (0.4–13)
Demers et al. 1994	K	Firefighters	SIR	0.9 (0.4–1.9)
Kang et al. 2008	K	Firefighters	SMOR	0.77 (0.31–1.92)
Burnett et al. 1994	S	Firefighters	PMR	*1.32 (1.02–1.67)*
Ma et al. 2006	S	Firefighters	SIR	1.09 (0.61–1.80)
Elliott et al. 1996	S	Incinerator communities	SMR	*1.03 (1.00–1.07)*
Floret et al. 2003	S	Incinerator communities	OR	*2.3 (1.4–3.8)*
Michelozzi et al. 1998	S	Incinerator communities	SMR	2.51 (0.29–8.51)
Viel et al. 2008	S	Incinerator communities	RR	*1.12 (1.002–1.251)*
Viel et al. 2000	S	Incinerator communities	SIR	*1.27 (p = 0.00003)*
Multiple Myeloma				
Aronson et al. 1994	K	Firefighters	SMR	0.47 (0.01–2.59)

TABLE 6-2 Continued

Study	Study Type	Population	Risk Estimate (95% CI)	
Baris et al. 2001	K	Firefighters	SMR	1.68 (0.90–3.11)
Bates et al. 2007	K	Firefighters	OR	1.03 (0.75–1.43)
Demers et al. 1994	K	Firefighters	SIR	0.7 (0.1–2.6)
Kang et al. 2008	K	Firefighters	SMOR	0.76 (0.39–1.48)
Burnett et al. 1994	S	Firefighters	PMR	***1.48 (1.02–2.07)***
Leukemia				
Aronson et al. 1994	K	Firefighters	SMR	1.20 (0.33–3.09)
Aronson et al. 1994	K	Firefighters	SMR	1.90 (0.52–4.88)
Baris et al. 2001	K	Firefighters	SMR	0.83 (0.50–1.37)
Bates et al. 2007	K	Firefighters	OR	1.22 (0.99–1.49)
Beaumont et al. 1991	K	Firefighters	SMR	0.61 (0.22–1.33)
Demers et al. 1994	K	Firefighters	SIR	1.0 (0.4–2.1)
Demers et al. 1994	K	Firefighters	IDR	0.8 (0.2–3.5)
Kang et al. 2008	K	Firefighters	SMOR	0.72 (0.43–1.20)
Bates et al. 2001	S	Firefighters	SIR	1.81 (0.5–4.6)
Burnett et al. 1994	S	Firefighters	PMR	1.19 (0.91–1.53)
Heyer et al. 1990	S	Firefighters	SMR	1.73 (0.70–3.58)
Ma et al. 2006	S	Firefighters	SIR	0.77 (0.47–1.19)
Michelozzi et al. 1998	S	Incinerator communities	SMR	0.82 (0.03–4.09)
Other				
Aronson et al. 1994	K	Firefighters	SMR	1.20 (0.33–3.09)
Beaumont et al. 1991	K	Firefighters	SMR	1.11 (0.76–1.58)
Deschamps et al. 1995	S	Firefighters	SMR	1.18 (0.14–4.27)

NOTE: Risk estimates in bold italics denote significant differences in risk. K = key study; S = supporting study.

^aThe risk estimates included in Table 6-2 reflect the cause-specific mortality or diagnosis for the study population as a whole. Many of these studies also include risk estimates by work duration, time since first exposure, or latency, but those data are not included here. For more detailed results, see chapter text or Appendix C.

better exposure assessment, the committee concluded there is no evidence for any one cancer site that rises above the level of inadequate/insufficient. Studies of Gulf War veterans exposed to oil-well–fire smoke also showed no cancer sites of concern, including brain cancer.

Based on a review of the epidemiologic literature, the committee concludes that there is *inadequate/insufficient* evidence of an association between long-term exposure to combustion products and cancer in the populations studied.

ALL-CAUSE MORTALITY

Cohort mortality studies often present SMRs for all causes of death combined as well as for specific causes of death. The all-cause SMR is an indication of how healthy the study population is in comparison to the population chosen to represent the background rates of disease. An all-cause SMR serves as a measure of the comparability of the study group with the reference population and can highlight potential biases in the study design and analysis.

All-Cause Mortality in Firefighters

Occupational cohort mortality studies comparing an employed group with the general population typically report SMRs for all causes of death combined as less than 1.0, indicating that fewer cohort members died than expected after adjusting for age, race, gender, and calendar year. One explanation for the commonly observed deficit is the healthy worker effect, a form of selection bias reflected in the better health status of workers relative to the general population (Fox and Collier 1976). The SMR for all causes of death combined can be interpreted

TABLE 6-3 All-Cause Mortality Among Firefighters

Study	SMR (95% CI)
Aronson et al. 1994	0.95 (0.88–1.02)
Baris et al. 2001	*0.96 (0.92–0.99)*[a]
Beaumont et al. 1991	*0.90 (0.85–0.95)*[a]
Demers et al. 1992a	*0.81 (0.77–0.86)*[a]
Deschamps et al. 1995	*0.52 (0.35–0.75)*[a]
Eliopulos et al. 1984	*0.80 (0.67–0.96)*[a]
Guidotti 1993	0.96 (0.87–1.07)
Hansen 1990	0.99 (0.75–1.29)
Heyer et al. 1990	*0.76 (0.69–0.85)*[a]
Ma et al. 2005	*0.57 (0.54–0.60)*[a]
Tornling et al. 1994	*0.82 (0.73–0.91)*[a]
Vena and Fiedler 1987	0.95 (0.87–1.04)

[a]SMR values in bold italics indicate a significant difference.

as a measure of the extent of the bias in a particular study (Monson 1986). There is some evidence that health worker effect bias affects chronic disease mortality such as cardiovascular, lung, and digestive diseases more than cancer (Blair et al. 1986).

Several of the studies reviewed earlier in this chapter report results for all-cause mortality. In particular, Vena and Fiedler (1987), Hansen (1990), Guidotti et al. (1993), and Aronson et al. (1994), reported SMRs of less than 1.0 with wide confidence intervals that included the null. Eight studies, Eliopulos et al. (1984), Heyer et al. (1990), Beaumont et al. (1991), Demers et al. (1992a), Tornling et al. (1994), Deschamps et al. (1995), Baris et al. (2001), and Ma et al. (2005), and found that overall mortality in firefighters was significantly lower than the reference groups. See Table 6-3 for the all-cause mortality SMRs for firefighters.

All-Cause Mortality Among Incinerator Workers and Communities Near Incinerators

One study (Gustavsson 1989) reported elevated all-cause mortality for incinerator workers compared to National Swedish mortality rates and to local rates for the greater Stockholm area (SMR 1.13, 95% CI 0.9–1.14, and SMR 0.99, 95% CI 0.79–1.22, respectively).

Conclusion

The committee concludes that there is *insufficient/inadequate* evidence to determine whether an association exists between all-cause mortality and exposure to combustion products in the populations studied.

SUMMARY

The committee studied the epidemiologic literature on exposure to combustion products from sources believed to be relevant to the burn pit exposures at JBB and other bases with burn pits in Iraq and Afghanistan. From that epidemiologic literature, the committee concludes that further study of health effects specifically among OEF/OIF veterans is necessary. The 2010 joint services report describing several health outcomes in military personnel at bases with and without operating burn pits is a first step in addressing some of these issues but the period of follow-up is too short to detect long-term health effects in this population (AFHSC et al. 2010). Further follow-up of these populations is warranted.

The research considered in this chapter is a best attempt to use currently available information on occupational and residential exposures to combustion products to extrapolate to exposures of military personnel stationed at JBB. However, because of differences in exposure parameters, population characteristics, access to medical care, and monitoring of health, the results for firefighters, incineration workers, and people living near incinerators might not be generalizable to military personnel exposed to burn pit emissions. Furthermore, although the difficulties in determining exposure to burn pits are apparent (see Chapter 4) such assessments are critical if adverse health outcomes that might result from such exposures are to be distinguished from those that might result from deployment per se and desert environments. The committee recognizes that the risks associated with being a firefighter, incinerator worker, or living near and incinerator might not provide a comprehensive picture of the risks posed to military personnel from burn pit emissions. Nevertheless, given the lack of information on the health effects associated with such exposures, the committee believes that studies of these surrogate populations provide a reasonable approach for evaluating the long-term consequences of exposure to combustion products similar to burn pit emissions.

Based on a review of the epidemiologic literature presented in this chapter, the committee concludes that there is limited/suggestive evidence of an association between exposure to combustion products and reduced pulmonary function in the populations studied. However, there is inadequate/insufficient evidence of an association between exposure to combustion products and cancer, respiratory disease, circulatory disease, neurologic disease, and adverse reproductive and developmental outcomes in the populations studied.

REFERENCES

AFHSC (U.S. Armed Forces Health Surveillance Center), the Naval Health Research Center, and the U.S. Army Public Health Command. 2010. Epidemiological Studies of Health Outcomes among Troops Deployed to Burn Pit Sites. Silver Spring, MD: Defense Technical Information Center. May.

Aronson, K. J., L. A. Dodds, L. Marrett, and C. Wall. 1996. Congenital anomalies among the offspring of fire fighters. *American Journal of Industrial Medicine* 30(1):83-86.

Aronson, K. J., G. A. Tomlinson, and L. Smith. 1994. Mortality among fire fighters in metropolitan Toronto. *American Journal of Industrial Medicine* 26(1):89-101.

Bandaranayke, D., D. Read, and Clare. Salmond. 1993. Health consequences of a chemical fire. *Internatinal Journal of Environmental Health Research* 3:104-114.

Baris, D., T. J. Garrity, J. L. Telles, E. F. Heineman, A. Olshan, and S. H. Zahm. 2001. Cohort mortality study of Philadelphia firefighters. *American Journal of Industrial Medicine* 39(5):463-476.

Barth, Shannon K., Han K. Kang, Tim A. Bullman, and Mitchell T. Wallin. 2009. Neurological mortality among U.S. veterans of the Persian Gulf War: 13-year follow-up. *American Journal of Industrial Medicine* 52(9):663-670.

Bartoo, Carole. 2010. Soldiers' mysterious lung disease identified. *The Reporter: Vanderbilt University Medical Center's Weekly Newspaper*, February 12.

Bates, J. T. 1987. Coronary artery disease deaths in the Toronto Fire Department. *Journal of Occupational Medicine* 29(2):132-135.

Bates, M. N. 2007. Registry-based case-control study of cancer in California firefighters. *American Journal of Industrial Medicine* 50(5):339-344.

Bates, M. N., J. Fawcett, N. Garrett, R. Arnold, N. Pearce, and A. Woodward. 2001. Is testicular cancer an occupational disease of fire fighters? *American Journal of Industrial Medicine* 40(3):263-270.

Beaumont, J. J., G. S. Chu, J. R. Jones, M. B. Schenker, J. A. Singleton, L. G. Piantanida, and M. Reiterman. 1991. An epidemiologic study of cancer and other causes of mortality in San Francisco firefighters. *American Journal of Industrial Medicine* 19(3):357-372.

Betchley, C., J. Q. Koenig, G. van Belle, H. Checkoway, and T. Reinhardt. 1997. Pulmonary function and respiratory symptoms in forest firefighters. *American Journal of Industrial Medicine* 31(5):503-509.

Biggeri, A., F. Barbone, C. Lagazio, M. Bovenzi, and G. Stanta. 1996. Air pollution and lung cancer in Trieste, Italy: Spatial analysis of risk as a function of distance from sources. *Environmental Health Perspectives* 104(7):750-754.

Blair, A., P. Stewart, M. O'Berg, W. Gaffey, J. Walrath, J. Ward, R. Bales, S. Kaplan, and D. Cubit. 1986. Mortality among industrial workers exposed to formaldehyde. *Journal of National Cancer Institute* 76(6):1071-1084.

Bresnitz, E. A., J. Roseman, D. Becker, and E. Gracely. 1992. Morbidity among municipal waste incinerator workers. *American Journal of Industrial Medicine* 22(3):363-378.

Bullman, T. A., C. M. Mahan, H. K. Kang, and W. F. Page. 2005. Mortality in US Army Gulf War veterans exposed to 1991 Khamisiyah chemical munitions destruction. *American Journal of Public Health* 95(8):1382-1388.

Burnett, C. A., W. E. Halperin, N. R. Lalich, and J. P. Sestito. 1994. Mortality among fire fighters: A 27 state survey. *American Journal of Industrial Medicine* 26(6):831-833.

Calvert, G. M., J. W. Merling, and C. A. Burnett. 1999. Ischemic heart disease mortality and occupation among 16- to 60-year-old males. *Journal of Occupational Environmental Medicine* 41(11):960-966.

Carozza, S. E., M. Wrensch, R. Miike, B. Newman, A. F. Olshan, D. A. Savitz, M. Yost, and M. Lee. 2000. Occupation and adult gliomas. *American Journal of Epidemiology* 152(9):838-846.

CDC (Centers for Disease Control and Prevention). 2009. *Heart Disease Risk Factors*. CDC, October 29. http://www.cdc.gov/heartdisease/risk_factors.htm (accessed January 13, 2011).

Charbotel, B., M. Hours, A. Perdrix, L. Anzivino-Viricel, and A. Bergeret. 2005. Respiratory function among waste incinerator workers. *International Archives of Occupational and Environmental Health* 78(1):65-70.

Chia, S. E., and L. M. Shi. 2002. Review of recent epidemiologic studies on paternal occupations and birth defects. *Occupational and Environmental Medicine* 59(3):149-155.

Choi, B. C. 2000. A technique to re-assess epidemiologic evidence in light of the healthy worker effect: The case of firefighting and heart disease. *Journal of Occupational & Environmental Medicine* 42(10):1021-1034.

Comba, P., V. Ascoli, S. Belli, M. Benedetti, L. Gatti, P. Ricci, and A. Tieghi. 2003. Risk of soft tissue sarcomas and residence in the neighbourhood of an incinerator of industrial wastes. *Occupational and Environmental Medicine* 60(9):680-683.

Cordier, S., C. Chevrier, E. Robert-Gnansia, C. Lorente, P. Brula, and M. Hours. 2004. Risk of congenital anomalies in the vicinity of municipal solid waste incinerators. *Occupational and Environmental Medicine* 61(1):8-15.

Cowan, D. N., J. L. Lange, J. Heller, J. Kirkpatrick, and S. DeBakey. 2002. A case-control study of asthma among U.S. Army Gulf War veterans and modeled exposure to oil well fire smoke. *Military Medicine* 167(9):777-782.

Cresswell, P. A., J. E. Scott, S. Pattenden, and M. Vrijheid. 2003. Risk of congenital anomalies near the Byker waste combustion plant. *Journal of Public Health Medicine* 25(3):237-242.

Demers, P. A., H. Checkoway, T. L. Vaughan, N. S. Weiss, N. J. Heyer, and L. Rosenstock. 1994. Cancer incidence among firefighters in Seattle and Tacoma, Washington (United States). *Cancer Causes Control* 5(2):129-135.

Demers, P. A., N. J. Heyer, and L. Rosenstock. 1992a. Mortality among firefighters from three northwestern United States cities. *British Journal of Industrial Medicine* 49(9):664-670.

Demers, P. A., T. L. Vaughan, et al. 1992b. Cancer identification using a tumor registry versus death certificates in occupational cohort studies in the United States. *American Journal of Epidemiology* 136(10):1232-1240.

Deschamps, S., I. Momas, and B. Festy. 1995. Mortality amongst Paris fire-fighters. *European Journal of Epidemiology* 11(6):643-646.

Dibbs, E., H. E. Thomas, S. T. Weiss, and D. Sparrow. 1982. Fire fighting and coronary heart disease. *Circulation* 65(5):943-946.

Douglas, D. B., R. B. Douglas, D. Oakes, and G. Scott. 1985. Pulmonary function of London firemen. *British Journal of Industrial Med* 42(1):55-58.

Elci, O. C., M. Akpinar-Elci, M. Alavanja, and M. Dosemeci. 2003. Occupation and the risk of lung cancer by histologic types and morphologic distribution: A case control study in Turkey. *Monaldi Archives for Chest Disease* 59(3):183-188.

Eliopulos, E., B. K. Armstrong, J. T. Spickett, and F. Heyworth. 1984. Mortality of fire fighters in Western Australia. *British Journal of Industrial Medicine* 41(2):183-187.

Elliott, P., G. Shaddick, I. Kleinschmidt, D. Jolley, P. Walls, J. Beresford, and C. Grundy. 1996. Cancer incidence near municipal solid waste incinerators in Great Britain. *British Journal of Cancer* 73(5):702-710.

Feuer, E., and K. Rosenman. 1986. Mortality in police and firefighters in New Jersey. *American Journal of Industrial Medicine* 9(6):517-527.

Firth, H. M., K. R. Cooke, and G. P. Herbison. 1996. Male cancer incidence by occupation: New Zealand, 1972-1984. *International Journal of Epidemiology* 25(1):14-21.

Floret, N., F. Mauny, B. Challier, P. Arveux, J. Y. Cahn, and J. F. Viel. 2003. Dioxin emissions from a solid waste incinerator and risk of non-Hodgkin lymphoma. *Epidemiology* 14(4):392-398.

Fox, A. J., and P. F. Collier. 1976. Low mortality rates in industrial cohort studies due to selection for work and survival in the industry. *British Journal of Preventive & Social Medicine* 30(4):225-230.

Glod, M. 2010. Alarms sound over trash fires in war zones of Afghanistan, Iraq. *The Washington Post* (August 6, 2010).

Glueck, C. J., W. Kelley, P. Wang, P. S. Gartside, D. Black, and T. Tracey. 1996. Risk ractors for coronary heart disease among firefighters in Cincinnati. *American Journal of Industrial Medicine* 30:331-340.

Golden, A. L., S. B. Markowitz, and P. J. Landrigan. 1995. The risk of cancer in firefighters. *Occupational Medicine* 10(4):803-820.

Grimes, G., D. Hirsch, and D. Borgeson. 1991. Risk of death among Honolulu fire fighters. *Hawaii Medical Journal* 50(3):82-85.

Guidotti, T. L. 1993. Mortality of urban firefighters in Alberta, 1927-1987. *American Journal of Industrial Medicine* 23(6):921-940.

Guidotti, T. L. 1995. Occupational mortality among firefighters: assessing the association. *Journal of Occupational Environmental Medicine* 37(12):1348-1356.

Gustavsson, P. 1989. Mortality among workers at a municipal waste incinerator. *American Journal of Industria Medicine* 15(3):245-253.

Gustavsson, P., B. Evanoff, and C. Hogstedt. 1993. Increased risk of esophageal cancer among workers exposed to combustion products. *Archives of Environmental Health* 48(4):243-245.

Haas, N. S., M. Gochfeld, M. G. Robson, and D. Wartenberg. 2003. Latent health effects in firefighters. *International Journal of Occupational Environmental Health* 9(2):95-103.

Hansen, E. S. 1990. A cohort study on the mortality of firefighters. *British Journal of Industrial Medicine* 47(12):805-809.

Hazucha, M. J., V. Rhodes, B. A. Boehlecke, K. Southwick, D. Degnan, and C. M. Shy. 2002. Characterization of spirometric function in residents of three comparison communities and of three communities located near waste incinerators in North Carolina. *Archives of Environmental Health* 57(2):103-112.

Heyer, N., N. S. Weiss, P. Demers, and L. Rosenstock. 1990. Cohort mortality study of Seattle fire fighters: 1945-1983. *American Journal of Industrial Medicine* 17(4):493-504.

Horsfield, K., F. M. Cooper, M. P. Buckman, A. R. Guyatt, and G. Cumming. 1988a. Respiratory symptoms in West Sussex firemen. *British Journal of Industrial Medicine* 45(4):251-255.

Horsfield, K., A. R. Guyatt, F. M. Cooper, M. P. Buckman, and G. Cumming. 1988b. Lung function in West Sussex firemen: A four year study. *British Journal of Industrial Medicine* 45(2):116-121.

Hours, M., L. Anzivino-Viricel, A. Maitre, A. Perdrix, Y. Perrodin, B. Charbotel, and A. Bergeret. 2003. Morbidity among municipal waste incinerator workers: A cross-sectional study. *International Archives of Occupational and Environmental Health* 76(6):467-472.

Howe, G. R., and J. D. Burch. 1990. Fire fighters and risk of cancer: An assessment and overview of the epidemiologic evidence. *American Journal of Epidemiology* 132(6):1039-1050.

Hu, S. W., and C. M. Shy. 2001. Health effects of waste incineration: a review of epidemiologic studies. *Journal of the Air & Waste Management Association* 51(7):1100-1109.

Hu, S. W., M. Hazucha, and C. M. Shy. 2001. Waste incineration and pulmonary function: An epidemiologic study of six communities. *Journal of the Air & Waste Management Association* 51(8):1185-1194.

IARC (International Agency on Research for Cancer). 2004. Tobacco smoke and involuntary smoking. *IARC Monographs on the Evaluation of Carcinogenic Risks to Humans*. Volume 83. Lyon, France: IARC Press.

IARC. 2010. Painting, firefighting, and shiftwork. *IARC Monographs on the Evaluation of Carcinogenic Risks to Humans*. Volume 98. Lyon, France: IARC Press.

IOM (Institute of Medicine). 2000. *Gulf War and health: Volume 1. Depleted uranium, pyridostigmine bromide, sarin, vaccines*. Washington, DC: National Academy Press.

IOM. 2005. *Gulf War and health: Volume 3. Fuels, combustion products, and propellants*. Washington, DC: The National Academies Press.

IOM. 2006. *Gulf War and health: Volume 4. Health effects of serving in the Gulf War*. Washington, DC: The National Academies Press.

IOM. 2009. *Combating tobacco in military and veteran populations*. Washington, DC: The National Academies Press.

IOM. 2010. *Gulf War and health: Volume 8. Update of health effects of serving in the Gulf War*. Washington, DC: The National Academies Press.

Iowa Persian Gulf Study Group. 1997. Self-reported illness and health status among Gulf War veterans: A population-based study. *JAMA Journal of the American Medical Association* 277(3):238-245.

Jansson, B., and L. Voog. 1989. Dioxin from Swedish municipal incinerators and the occurrence of cleft lip and palate malformations. *International Journal of Environmental Studies* 34:99-104.

Kang, D., L. K. Davis, P. Hunt, and D. Kriebel. 2008. Cancer incidence among male Massachusetts firefighters, 1987-2003. *American Journal of Industrial Medicine* 51(5):329-335.

Kang, H. K., C. M. Mahan, K. Y. Lee, C. A. Magee, and F. M. Murphy. 2000. Illnesses among United States veterans of the Gulf War: A population-based survey of 30,000 veterans. *Journal of Occupational and Environmental Medicine* 42(5):491-501.

Kilburn, K. H., R. H. Warsaw, and M. G. Shields. 1989. Neurobehavioral dysfunction in firemen exposed to polycholorinated biphenyls (PCBs): Possible improvement after detoxification. *Archives of Environmental Health* 44(6):345-350.

King, M. S., R. Eisenberg, J. H. Newman, J. J. Tolle, F. E. Harrell, Jr., H. Nian, M. Ninan, E. S. Lambright, J. R. Sheller, J. E. Johnson, and R. F. Miller. 2011. Constrictive bronchiolitis in soldiers returning from Iraq and Afghanistan. *New England Journal of Medicine* 365(3):222-230.

King, M. S., R. Miller, J. Johnson, M. Ninan, E. Lambright, A. F. Shorr, and J. Sheller. 2008. Bronchiolitis in soldiers with inhalation exposures in the Iraq War. Poster presented at the *American Thoracic Society*'s 104th International Conference, Toronto, Canada, May 2008.

Krishnan, G., M. Felini, S. E. Carozza, R. Miike, T. Chew, and M. Wrensch. 2003. Occupation and adult gliomas in the San Francisco Bay Area. *Journal of Occupational and Environmental Medicine* 45(6):639-647.

Lange, J. L., D. A. Schwartz, B. N. Doebbeling, J. M. Heller, and P. S. Thorne. 2002. Exposures to the Kuwait oil fires and their association with asthma and bronchitis among Gulf War veterans. *Environmental Health Perspectives* 110(11):1141-1146.

Lee, J. T., and C. M. Shy. 1999. Respiratory function as measured by peak expiratory flow rate and PM_{10}: six communities study. *Journal of Exposure Analysis and Environmental Epidemiology* 9(4):293-299.

LeMasters, G. K., A. M. Genaidy, P. Succop, J. Deddens, T. Sobeih, H. Barriera-Viruet, K. Dunning, and J. Lockey. 2006. Cancer risk among firefighters: A review and meta-analysis of 32 studies. *Journal of Occupational and Environmental Medicine* 48(11):1189-1202.

Liu, D., I. B. Tager, J. R. Balmes, and R. J. Harrison. 1992. The effect of smoke inhalation on lung function and airway responsiveness in wildland fire fighters. *American Review of Respiratory Disease* 146(6):1469-1473.

Lloyd, O. L., M. M. Lloyd, F. L. Williams, and A. Lawson. 1988. Twinning in human populations and in cattle exposed to air pollution from incinerators. *British Journal of Industrial Medicine* 45(8):556-560.

Ma, F., L. E. Fleming, D. J. Lee, E. Trapido, and T. A. Gerace. 2006. Cancer incidence in Florida professional firefighters, 1981 to 1999. *Journal of Occupational and Environmental Medicine* 48(9):883-888.

Ma, F., L. E. Fleming, D. J. Lee, E. Trapido, T. A. Gerace, H. Lai, and S. Lai. 2005. Mortality in Florida professional firefighters, 1972 to 1999. *American Journal of Industrial Medicine* 47(6):509-517.

Ma, F., D. J. Lee, L. E. Fleming, and M. Dosemeci. 1998. Race-specific cancer mortality in U.S. firefighters: 1984-1993. *Journal of Occupational and Environmental Medicine* 40(12):1134-1138.

Markowitz, J. S. 1989. Self-reported short- and long-term respiratory effects among PVC-exposed firefighters. *Archives of Environmental Health* 44(1):30-33.

McDiarmid, M. A., and J. Agnew. 1995. Reproductive hazards and firefighters. *Occupational Medicine* 10(4):829-841.

McGregor, D. B. 2005. *Risk of brain tumours in firemen.* Studies and research projects report, R-397. Montreal, Quebec: Institut de Recherche Robert Sauvé en Santé et en Sécurité du Travail.

Medline Plus. 2011. *Pulmonary function testing.* Department of Health and Human Services, National Institutes of Health, February 28, 2011. http://www.nlm.nih.gov/medlineplus/ency/article/003853.htm (accessed March 1, 2011).

Michelozzi, P., F. Forastiere, D. Fusco, C. A. Perucci, B. Ostro, C. Ancona, and G. Pallotti. 1998. Air pollution and daily mortality in Rome, Italy. *Occupational and Environmental Medicine* 55(9):605-610.

Miedinger, D., P. N. Chhajed, D. Stolz, C. Gysin, A. B. Wanzenried, C. Schindler, C. Surber, H. C. Bucher, M. Tamm, and J. D. Leuppi. 2007. Respiratory symptoms, atopy and bronchial hyperreactivity in professional firefighters. *European Respiratory Journal* 30(3):538-544.

Milham, S., and E. Ossiander. 2001. Occupational Mortality Database. Washington State Department of Health. https://fortress.wa.gov/doh/occmort/OMQuery.aspx (accessed January 19, 2011).

Miller, R. 2009. Airway Injury in US Soldiers Following Service in Iraq and Afghanistan. In *United States Senate Committee on Veterans' Affairs Hearing.*

Mohan, A., D. Degnan, C. E. Feigley, C. M. Shy, C. A. Hornung, T. Mustafa, and C. A. Macera. 2000. Comparison of respiratory symptoms among community residents near waste disposal incinerators. *International Journal of Environmental Health Research* 10:63-75.

Monson, R. R. 1986. Observations on the healthy worker effect. *Journal of Occupational Medicine* 28(6):425-43.

Musk, A. W., R. R. Monson, J. M. Peters, and R. K. Peters. 1978. Mortality among Boston firefighters, 1915-1975. *British Journal of Industrial Medicine* 35(2):104-108.

Musk, A. W., J. M. Peters, L. Bernstein, C. Rubin, and C. B. Monroe. 1982. Pulmonary function in firefighters: A six-year follow-up in the Boston Fire Department. *American Journal of Industrial Medicine* 3(1):3-9.

Mustacchi, P. 1991. Neurobehavioral dysfunction in firemen exposed to polychlorinated biphenyls (PCBs): Possible improvement after detoxification. *Archives of Environmental Health* 46(4):254-255.

Mustajbegovic, J., E. Zuskin, E. N. Schachter, J. Kern, M. Vrcic-Keglevic, S. Heimer, K. Vitale, and T. Nada. 2001. Respiratory function in active firefighters. *American Journal of Industrial Medicine* 40(1):55-62.

NIEHS (National Institute of Environmental Health Sciences). 2011. *Cancer and the environment: What you need to know, what you can do.* NIH Publication No. 03-2039. http://www.niehs.nih.gov/about/materials/cancer-enviro.pdf (accessed March 1, 2011).

Olshan, A. F., K. Teschke, and P. A. Baird. 1990. Birth defects among offspring of firemen. *American Journal of Epidemiology* 131(2):312-321.

Peters, J. M., G. P. Theriault, L. J. Fine, and D. H. Wegman. 1974. Chronic effect of fire fighting on pulmonary function. *New England Journal of Medicine* 291(25):1320-1322.

Prezant, D. J., A. Dhala, A. Goldstein, D. Janus, F. Ortiz, T. K. Aldrich, and K. J. Kelly. 1999. The incidence, prevalence, and severity of sarcoidosis in New York City firefighters. *Chest* 116(5):1183-1193.

Proctor, S. P., T. Heeren, R. F. White, J. Wolfe, M. S. Borgos, J. D. Davis, L. Pepper, R. Clapp, P. B. Sutker, J. J. Vasterling, and D. Ozonoff. 1998. Health status of Persian Gulf War veterans: Self-reported symptoms, environmental exposures and the effect of stress. *International Journal of Epidemiology* 27(6):1000-1010.

Risen, James. 2010. Veterans sound alarm over burn-pit exposure. *The New York Times* (August 2, 2010).

Rosenstock, L., P. Demers, N. J. Heyer, and S. Barnhart. 1990. Respiratory mortality among firefighters. *British Journal of Industrial Medicine* 47(7):462-465.

Rydhstroem, H. 1998. No obvious spatial clustering of twin births in Sweden between 1973 and 1990. *Environmental Research* 76(1):27-31.

Sama, S. R., T. R. Martin, L. K. Davis, and D. Kriebel. 1990. Cancer incidence among Massachusetts firefighters, 1982-1986. *American Journal of Industrial Medicine* 18(1):47-54.

Sardinas, A., J. W. Miller, and H. Hansen. 1986. Ischemic heart disease mortality of firemen and policemen. *American Journal of Public Health* 76(9):1140-1141.

Schnitzer, P. G., A. F. Olshan, and J. D. Erickson. 1995. Paternal occupation and risk of birth defects in offspring. *Epidemiology* 6(6):577-583.

Serra, A., F. Mocci, and F. S. Randaccio. 1996. Pulmonary function in Sardinian fire fighters. *American Journal of Industrial Medicine* 30(1):78-82.

Shy, C. M., D. Degnan, D. I. Fox, S. Mukerjee, M. J. Hazucha, B. A. Boehlecke, D. Rothenbacher, P. M. Briggs, R. B. Devlin, D. D. Wallace, R. K. Stevens, and P. A. Bromberg. 1995. Do waste incinerators induce adverse respiratory effects? An air quality and epidemiologic study of six communities. *Environmental Health Perspectives* 103(7-8):714-724.

Smith, T. C., J. M. Heller, T. I. Hooper, G. D. Gackstetter, G. C. Gray. 2002. Are Gulf War veterans experiencing illness due to exposure to smoke from Kuwaiti oil well fires? Examination of Department of Defense hospitalization data. *American Journal of Epidemiology* 155 (10):908-917.

Sparrow, D., R. Bosse, B. Rosner, and S. T. Weiss. 1982. The effect of occupational exposure on pulmonary function: a longitudinal evaluation of fire fighters and nonfire fighters. *American Review of Respiratory Disease* 125(3):319-322.

Spencer, P. S., L. A. McCauley, J. A. Lapidus, M. Lasarev, S. K. Joos, and D. Storzbach. 2001. Self-reported exposures and their association with unexplained illness in a population-based case-control study of Gulf War veterans. *Journal of Occupational & Environmental Medicine* 43(12):1041-1056.

Szema, A. M, M. C. Peters, K. M. Weissinger, C. A. Gagliano, J. J. Chen. 2010. New-onset asthma among soldiers serving in Iraq and Afghanistan. *Allergy and Asthma Procedings* 31:1-5.

Tango, T., T. Fujita, T. Tanihata, M. Minowa, Y. Doi, N. Kato, S. Kunikane, I. Uchiyama, M. Tanaka, and T. Uehata. 2004. Risk of adverse reproductive outcomes associated with proximity to municipal solid waste incinerators with high dioxin emission levels in Japan. *Journal of Epidemiology* 14(3):83-93.

Taylor, G., V. Rush, A. Peck, and J. A. Vietas. 2008. *Screening health risk assessment burn pit exposures, JBB Air Base, Iraq, and addendum report.* IOH-RS-BR-TR-2008-0001/USACHPPM 47-MA-08PV-08. Brooks City-Base, TX: Air Force Institute for Operational Health and U.S. Army Center for Health Promotion and Preventative Medicine.

Tepper, A., G. W. Comstock, and M. Levine. 1991. A longitudinal study of pulmonary function in fire fighters. *American Journal of Industrial Medicine* 20(3):307-316.

Tornling, G., P. Gustavsson, and C. Hogstedt. 1994. Mortality and cancer incidence in Stockholm fire fighters. *American Journal of Industrial Medicine* 25(2):219-228.

Unger, K. M., R. M. Snow, J. M. Mestas, and W. C. Miller. 1980. Smoke inhalation in firemen. *Thorax* 35 (11):838-842.

Unwin, C., N. Blatchley, W. Coker, S. Ferry, M. Hotopf, L. Hull, K. Ismail, I. Palmer, A. David, and S. Wessely. 1999. Health of UK servicemen who served in Persian Gulf War. *Lancet* 353(9148):169-178.

U.S. Surgeon General. 1964. *Smoking and health: Report of the advisory committee to the surgeon general of the Public Health Service.* Public Health Service publication no. 1103. Washington, DC: U.S. Department of Health, Education, and Welfare, Public Health Service.

U.S. Surgeon General. 2004. *The health consequences of smoking: A report of the surgeon general.* Washington, DC: U.S. Department of Health and Human Services, Public Health Service.

USAPHC (U.S. Army Public Health Command). 2010. *Epidemiological Consultation No. 64-FF-064C-07, Mishraq Sulfur Fire Environmental Exposure Assessment, June 2003–March 2007.* Aberdeen Proving Ground, MD: U.S. Army Public Health Command. June.

Vena, J. E., and R. C. Fiedler. 1987. Mortality of a municipal-worker cohort: IV. Fire fighters. *American Journal of Industrial Medicine* 11(6):671-684.

Verret, C., M. A. Jutand, C. De Vigan, M. Begassat, L. Bensefa-Colas, P. Brochard, and R. Salamon. 2008. Reproductive health and pregnancy outcomes among French Gulf War veterans. *Biomed Central Public Health* 8:141.

Viel, J. F., P. Arveux, J. Baverel, and J. Y. Cahn. 2000. Soft-tissue sarcoma and non-Hodgkin's lymphoma clusters around a municipal solid waste incinerator with high dioxin emission levels. *American Journal of Epidemiology* 152(1):13-19.

Viel, J. F., C. Daniau, S. Goria, P. Fabre, P. de Crouy-Chanel, E. A. Sauleau, and P. Empereur-Bissonnet. 2008. Risk for non-Hodgkin's lymphoma in the vicinity of French municipal solid waste incinerators. *Environmental Health: A Global Access Science Source* 7:51.

Vinceti, M., C. Malagoli, S. Teggi, S. Fabbi, C. Goldoni, G. De Girolamo, P. Ferrari, G. Astolfi, F. Rivieri, and M. Bergomi. 2008. Adverse pregnancy outcomes in a population exposed to the emissions of a municipal waste incinerator. *Science of the Total Environment* 407(1):116-121.

White, R. F., S. P. Proctor, T. Heeren, J. Wolfe, M. Krengel, J. Vasterling, K. Lindem, K. J. Heaton, P. Sutker, and D. M. Ozonoff. 2001. Neuropsychological function in Gulf War veterans: Relationships to self-reported toxicant exposures. *American Journal of Industrial Medicine* 40(1):42-54.

Williams, F. L., A. B. Lawson, and O. L. Lloyd. 1992. Low sex ratios of births in areas at risk from air pollution from incinerators, as shown by geographical analysis and 3-dimensional mapping. *International Journal of Epidemiology* 21(2):311-319.

Wolfe, J., S. P. Proctor, D. J. Erickson, and H. Hu. 2002. Risk factors for multisymptom illness in U.S. Army veterans of the Gulf War. *Journal of Occupational & Environmental Medicine* 44(3):271-281.

Youakim, S. 2006. Risk of cancer among firefighters: a quantitative review of selected malignancies. *Archives of Environmental and Occupational Health* 61(5):223-231.

Young, I., J. Jackson, and S. West. 1980. Chronic respiratory disease and respiratory function in a group of fire fighters. *Medical Journal of Australia* 1(13):654-658.

Zambon, P., P. Ricci, E. Bovo, A. Casula, M. Gattolin, A. R. Fiore, F. Chiosi, and S. Guzzinati. 2007. Sarcoma risk and dioxin emissions from incinerators and industrial plants: A population-based case-control study (Italy). *Environmental Health: A Global Access Science Source* 6:19.

7

Synthesis and Conclusions

KEY FINDINGS

This report highlights findings and analyses in two broad categories: potential exposure to burn pit emissions at Joint Base Balad (JBB) and other U.S. military bases in Iraq and Afghanistan and potential health effects of that exposure. From the starting points outlined in Figure 3-2, the committee's data collection and analysis led to the expanded and interactive processes shown in Figure 7-1. Those processes draw on several information sources, such as air monitoring data from JBB, results of health-effects studies in military reports, and results of health-effects studies of related exposure scenarios in the medical and scientific literature, including literature on health effects associated with combustion products individually and in the aggregate. Although those sources yielded substantial amounts of information, the available database is marked by numerous data gaps and uncertainties that point to the need for additional studies and analysis, particularly an epidemiologic study of populations exposed to burn pit emissions. The committee summarizes below its findings presented in the previous chapters and synthesizes its findings into an assessment of the potential long-term health consequences for military and other personnel stationed at JBB and other sites that have burn pits.

Burn Pit Basics

Background information on the use of military burn pits was available from several sources, including the Department of Defense (DoD), particularly the U.S. Army Center for Health Promotion and Preventive Medicine (CHPPM, now the U.S. Army Public Health Command), the Government Accountability Office, and the RAND Corporation. In spite of air-monitoring data from JBB, incomplete information on the burning practices and on the materials burned in the pits hampered the committee's ability to determine the chemical composition and amount of emissions from the burn pits at JBB and elsewhere. For example, burning of tires was banned at some bases and apparently permitted at others. The committee asked the DoD for specific information pertaining to burn pit locations on the bases and in Iraq and Afghanistan, the number of burn pits in use by the military in those countries, their frequency of use, and the average burn times for the pits. The DoD was unable to provide specific information to characterize waste streams at large and small forward operating bases but it did provide generic data on waste-stream content, for example, percentages of plastics, wood, and miscellaneous noncombustible, metal, and combustible materials at bases in Iraq and Afghanistan (Faulkner 2011). Nonhazardous solid waste generated by bases in Kosovo, Bosnia, and Bulgaria was better characterized (Faulkner 2011), but the extent of similarity

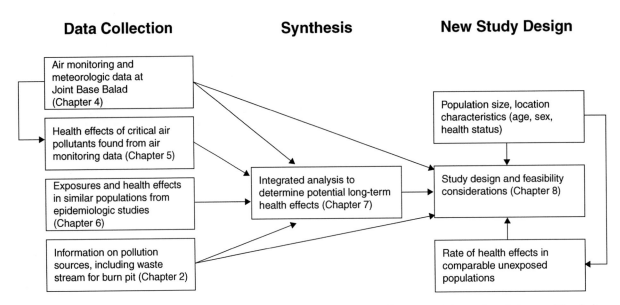

FIGURE 7-1 Committee's process for assessing the long-term consequences of exposure to burn pit emissions and the design of an epidemiologic study.

between the burn pits in these countries and in Iraq and Afghanistan, and in particular at JBB, is unknown. Available information suggests that burn pit practices varied among bases and over time; the amount of waste burned daily at JBB is estimated to have varied by an order of magnitude over its years of operation.

Combustion Products and Exposure Considerations

Air samples were collected upwind of the burn pit and downwind in an open area and in a housing and work area. The sampling targeted several chemical categories, including particulate matter (PM) and metals, volatile organic compounds (VOCs), polycyclic aromatic hydrocarbons (PAHs), and polychlorinated dibenzodioxins and furans (PCDDs/Fs). It is notable that some U.S. criteria air pollutants, such as sulfur dioxide, nitrogen dioxide, carbon monoxide, and ozone, were not analyzed for at JBB. CHPPM found measurable concentrations of many of the chemicals for which the samples were analyzed, albeit at concentrations below those typically observed in urban airsheds outside the United States. However, the samples contained contributions from other sources of air pollution in the area, so measured air contaminants could not be directly attributed to the JBB burn pit. Based on its analysis of the CHPPM data, the committee concluded that:

- Background ambient air concentrations of PM at JBB are high, with average concentrations above the U.S. air pollution standards, and are most likely derived from local sources;
- PCDDs/Fs were detected at low concentrations in nearly all samples (although high compared with even polluted urban areas), and the burn pit was likely the major source of these chemicals; and
- Ambient VOC and PAH concentrations were similar to those reported for polluted urban environments outside the United States, and the major sources of these pollutants are regional background, ground transportation, stationary power generation, and the JBB airport.

The data thus indicate that military personnel were exposed to a mixture of combustion products from the burn pit and to other air pollutants from local and regional sources, including other combustion sources, industry, and wind-blown soil. It is likely that people who worked at or near burn pits were exposed to combustion emissions and products of incomplete combustion at concentrations higher than those measured at the CHPPM monitoring sites, but the available information is insufficient to estimate the concentrations to which they might have been exposed.

Potentially Related Health Effects and Hazard Considerations

Most of the pollutants detected at JBB were present at concentrations lower than health-based reference values of individual pollutants that are considered to be protective of the general U.S. population. The exceptions were PM and acrolein, neither of which could be attributed primarily to the burn pit. Measured PM concentrations substantially exceeded U.S. standards, and increased cardiovascular and respiratory morbidity and mortality have been observed in many epidemiologic studies of PM exposures at lower concentrations (IOM 2010). Although acrolein was poorly measured with the analytic techniques used, the concentrations were above a health-based reference concentration of 2×10^{-5} mg/m^3 (EPA 2003) although still far below any concentration observed to cause the nasal lesions and lung damage in laboratory animal experiments from which the reference value is derived. Exceedances of the health-based reference value of acrolein are common in U.S. monitoring data.

However, information on health effects of exposure to mixtures specific to burn pits is minimal. There is a larger dataset for health effects of exposure to combustion products from the open burning of waste and of exposure to combustion products from the burning of cooking and heating fuels, such vehicle fuels as JP-8, and household and municipal waste. The committee therefore turned to studies of firefighters and workers at municipal incinerators as populations exposed to combustion products likely to be similar in chemical composition to burn pit emissions. It was hoped that those populations would provide information useful for assessing possible long-term health effects of exposure to such mixtures.

That approach was based on several considerations. The DoD had labeled the waste stream for the military burn pits as containing "household waste," what is known of its composition is similar to that of such waste, and the dioxin congener profiles measured at JBB are similar to those seen in open burning of household waste. Firefighters are exposed to burning homes, commercial properties, and wildfires that might also contain materials similar to burn pit material, such as wood, electronics, and particularly plastics. Municipal waste-incinerator workers would be expected to be exposed to similar combustion products. Because the military personnel at JBB not only worked on the base but lived there, the committee also sought information on reported long-term health effects in residents who live near municipal waste incinerators.

Those studies did not yield conclusive results about long-term health effects associated with exposure to combustion products in the populations studied, but several health outcomes deserve further investigation. Respiratory outcomes should be assessed further because such air pollutants as PM, which are generally present in burn pit emissions, are known to have respiratory effects. Pulmonary function should be monitored to track changes and diagnose respiratory diseases, including asthma, bronchitis, and emphysema. Although there are reports of constrictive bronchiolitis in military personnel returning from JBB, it is difficult to conduct studies to assess this effect because the diagnosis is based on lung biopsy and ascertainment of cases for followup studies is therefore problematic. Nevertheless, the presentation of several cases in military personnel who have been deployed to bases that have burn pits warrants continuing study.

Many neurologic outcomes are associated with chemicals potentially present in emissions from burn pits. However, studies involving firefighters and others exposed to incinerator emissions provide little information on neurologic outcomes that might be associated with exposure to mixtures of combustion products.

Cardiovascular outcomes were observed in some studies; however, no increase in any specific cardiovascular outcome was reported consistently. Increased ischemic heart disease, arteriosclerosis, aortic aneurysm, and diseases of arteries and veins were reported in studies of firefighters but not in studies of other populations. Most of the studies of firefighters and other populations lacked adjustment for the acknowledged risk factors of cardiovascular disease, such as diet, physical activity, smoking, and a family history of heart disease. Nevertheless, although the evidence provided by those epidemiologic studies did not allow the committee to reach definitive conclusions about the risk of cardiovascular disease, an increase in risk caused by exposure to combustion products cannot be ruled out. Future followup pertaining to cardiovascular disease should develop data on risk factors, such as smoking, and should investigate disease incidence rather than focus on mortality. Better exposure assessment is critical.

The epidemiologic literature on cancer in firefighters and incinerator workers is inconsistent; reports link work in these occupations with brain, testicular, colon, and prostate cancers. Because many of the studies are marked by failure to control for confounding factors, have infrequent or inadequate exposure assessment, and do not have sufficient information to show dose–response relationships, the committee was unable to determine whether any increased risks of brain, testicular, colon, or prostate cancer were associated with combustion products or with other characteristics specific to the study populations. Because of the carcinogenic nature of many of the chemicals potentially associated with burn pit emissions, it is prudent to continue investigations of cancer end points and other health outcomes that have long latency in exposed military populations.

The DoD or the Department of Veterans Affairs (VA) may also consider investigating other health outcomes seen in personnel deployed to JBB, such as birth defects, rheumatoid arthritis, and lupus.

SYNTHESIS AND IMPLICATIONS

The starting point for the committee's work was a series of military reports and individual observations related to combustion products and possible health effects associated with military service at JBB, mainly from 2006 to 2009. Those reports and observations, which include videos and photographs, clearly establish that smoke from the burn pits drifted over the base at various times. However, the absence of atmospheric sampling data describing more fully the nature and extent of exposure to smoke by military personnel limits the availability of critical information, such as the concentration, composition, frequency, or duration of smoke episodes at the site. Similarly, although medical data on the health status of military personnel who served at JBB are available, the epidemiologic studies that have been conducted by the DoD do not provide sufficient information for evaluating an any association between exposure to burn pit emissions and long-term health effects in exposed military personnel. The DoD studies do not have sufficient followup to identify latent health effects, such as cancer and cardiovascular disease. The studies also do not provide any information on the degree of exposure that people may have experienced at JBB, such as how long they were at the base, where on the base they worked or lived, and whether they had more than one deployment to JBB or to other bases that had operating burn pits; nor do the studies adjust for confounding factors, such as smoking. The committee recognizes that the DoD studies did categorized deployed personnel as being with a 3- or 5-mile radius at some bases and compared them with deployed personnel stationed at bases without burn pits or with nondeployed personnel. The major limitations that the committee found in characterizing the waste stream at JBB and in the air monitoring data collected there and in assessing exposure and likely health effects of exposure to the burn pit emissions are given in Box 7-1.

Although air monitoring was conducted twice at JBB in 2007 and once in 2009, the locations of the sampling sites were not optimal for determining ambient concentrations of the air pollutants to which residents of the base might be exposed, and did not target smoke episodes. Only PM and acrolein were detected in the air samples at JBB at concentrations above reference concentrations intended for application for the general population in the United States. The committee understood that the limitations in the JBB air monitoring data would affect its work and that developing the analyses requested in the statement of task would require data from nonmilitary sources in addition to those collected previously at JBB and other military sites. These other sources, although useful, also had substantial limitations and did not permit the committee to determine precisely all the chemicals that might be in the burn pit emissions or their concentrations.

Nevertheless, the committee developed useful information on three categories of combustion products—PCDDs/Fs, VOCs and PAHs, and PM—and on their association with long-term health effects. The observed smoke from burn pits undoubtedly implies a contribution to the PM at JBB, but the committee was unable to distinguish any effects of the burn pit emissions on measured PM. The high PM concentrations come primarily from local sources, such as traffic and jet emissions, and from regional sources, including long-range anthropogenic emissions and dust storms. Health effects associated with PM exposure include cardiovascular and respiratory morbidity and mortality. Chronic cardiovascular and respiratory effects have been associated with PM concentrations substantially below those found at JBB, and this suggests that continued health monitoring of military personnel deployed to JBB would be appropriate.

On the basis of the CHPPM air monitoring data from JBB, PCDDs/Fs were detected at low concentrations

> **BOX 7-1**
> **Limitations of and Uncertainties in Burn Pit Information**
>
> - Waste stream characterization unknowns
>   - Content
>   - Volume
>   - Burn rate
>   - Variability
>   - Specific location and dimensions of pit
>
> - Air monitoring and meteorologic data
>   - Lack of monitoring for criteria air pollutants
>   - Meteorologic data not location-specific (not linked to sampling sites)
>   - Lack of specificity of data (no longitude and latitude for sampling sites or burn pits, airport, all housing areas, and local sources of pollutants)
>   - No monitoring conducted close to burn pit, so no direct sampling of burn pit emissions
>   - Lack of simultaneous sampling at all sites
>   - Lack of coordination between emissions and burn pit activities, particularly for smoke events
>
> - Exposure information
>   - Lack of information on personnel at JBB (such as how many people worked in or near the pits, for how long, and how frequently)
>   - No characterization of where people lived, ate, and worked
>   - Highly variable characterization of how long people were at JBB
>   - Other sources of exposure to air pollutants (such as from kerosene heaters, cooking appliances, and airport activities)
>   - No information on possible ingestion or dermal exposures (no soil-surface samples or indoor air samples)
>
> - Health effects
>   - Lack of characterization of PM to permit identification of possible health effects
>   - Use of surrogate populations and many uncertainties comparing exposures, duration, and frequency
>   - Surrogate-population studies lack dose–response information and exposure characterization
>   - DOD epidemiologic studies:
>   - Insufficient time for followup
>   - Looked at some relatively rare diseases for this population for long-term effects, such as birth defects, chronic multisymptom illness, lupus, and rheumatoid arthritis

in all samples, and the JBB burn pit was probably the major source of these chemicals, to judge by the measured gradient of concentrations with distance from the burn pit. The toxic equivalents of the dioxin concentrations are high compared with locations in the United States and are higher than those found in many polluted urban environments outside the United States. Health effects associated with exposure (of higher magnitudes) to dioxins and dioxin-like compounds include cancer, diabetes and other endocrine system effects, immunologic responses, neurologic effects, and reproductive and developmental effects.

The major sources of VOCs and PAHs at JBB were regional background, ground transportation, stationary power generation, and the airport on the base. The presence of these chemicals was not demonstrably associated with the burn pit, although the burn pit probably contributed to the ambient concentrations. VOC and PAH concentrations in the CHPPM monitoring study were similar to those reported in polluted urban areas outside the United

States. The measured concentrations were, in general and chemical by chemical, well below those expected to cause health effects in a population of healthy people. However, the effect of the overall mixture is not predictable with current methods.

The committee focused its literature review on long-term health outcomes in populations that were considered by the committee to have exposure to chemical mixtures that were similar to burn pit emissions: firefighters, municipal incinerator workers, and residents who lived near municipal incinerators. Although there was a wealth of epidemiologic information on health outcomes seen in firefighters, it was unclear how applicable it was to the military personnel at JBB and other military bases that had burn pits. The committee considered the risk potential for long-term health outcomes associated with combustion products in those surrogate populations and found limited or suggestive evidence of an association only for decreased pulmonary function in firefighters. Such a categorization warrants further study to develop the evidence base and support a reassessment of the strength of the association. There was insufficient or inadequate evidence of an association between exposure to combustion products and any other health effects in those surrogate populations. For all the other outcomes, the committee found inadequate or insufficient evidence of an association between long-term exposure to combustion products and specific health outcomes in the populations studied. That category can be used for a number of situations, for example, when the evidence is contradictory—that is, some well-conducted studies found an association but others did not—or when there are no key studies on which to base a conclusion and the supporting studies have substantial limitations.

As a result, the committee is unable to say whether exposures to emissions from the burn pit at JBB have caused long-term health effects. In particular, as summarized below, none of the individual chemical constituents among the combustion products emitted by the burn pit at JBB and measured in the monitoring campaigns appears to have been present at concentrations likely to be responsible for the adverse health outcomes studied in this report. However, the committee's review of the literature and the data from JBB suggests that service in Iraq or Afghanistan—that is, a broader consideration of air pollution than exposure only to burn pit emissions—might be associated with long-term health effects, particularly in susceptible (for example, those who have asthma) or highly exposed subpopulations (such as those who worked at the burn pit). Such health effects would be due mainly to high ambient concentrations of PM from both natural and anthropogenic sources, including military sources. If that broader exposure to air pollution turns out to be relevant, potentially related health effects of concern are respiratory and cardiovascular effects and cancer. Susceptibility to the PM health effects could be exacerbated by other exposures, such as stress, smoking, local climatic conditions, and coexposures to other chemicals that affect the same biologic or chemical processes.

Individually, the chemicals measured at JBB were generally below concentrations of health concern for general populations in the United States. However, the possibility of exposure to mixtures of the chemicals raises the potential for health outcomes associated with cumulative exposure to combinations of the constituents of burn pit emissions and emissions from other sources. As a preliminary step towards understanding possible long-term health effects of multiple contaminants or cumulative exposures, the committee looked at all the detected pollutants and the target organs or specific effects associated with them. The rationale for that approach is that when multiple pollutants contribute to a common outcome, the outcome may be more likely to occur, particularly if the pollutants have the same mechanisms of action. Many of the chemicals detected at JBB have the same toxic end points; for example, several are associated with liver toxicity or respiratory effects. Some of the health effects associated with five or more of the chemicals detected at JBB include some cancers, liver toxicity and reduced liver function, kidney toxicity and reduced kidney function, respiratory toxicity and morbidity, neurologic effects, blood effects (anemia and changes in various blood-cell types), cardiovascular toxicity and morbidity, and reproductive toxicity.

As summarized above, the studies reviewed for this report provide information about chemical agents found in burn pit emissions, some at very low concentrations and some at higher concentrations, and about related health outcomes, some of little potential concern and some of greater potential concern. But the limitations and uncertainties associated with the studies cannot be overlooked. In particular, the available epidemiologic studies are inconsistent in quality, were conducted with various degrees of methodologic rigor, and had considerable variations in design and sample size. Most critical, the database on the nature and extent of exposure to combustion products is incomplete.

Those considerations should be instructive for any future epidemiologic studies of JBB and other burn pit populations or other military personnel serving in Iraq and Afghanistan. In that regard, with awareness of the data gaps and analytic limitations in the studies reviewed here, the committee recommends that the DoD and the VA give special attention to several important aspects of the studies: uncertainties and limitations in exposure assessment, the scarcity of data on the long-term health effects resulting from exposure to relevant combustion products, and differences between the best surrogate populations described in the epidemiologic literature (firefighters and municipal incinerator workers) and military personnel stationed at JBB. Related feasibility and design elements for a future epidemiologic study are discussed in Chapter 8.

REFERENCES

EPA (U.S. Environmental Protection Agency). 2003. *Toxicological profile for acrolein.* CASRN 107-02-8. Integrated Risk Information System (IRIS) database. http://www.epa.gov/iris/subst/0364.htm (accessed July 12, 2011).

Faulkner, W.M. 2011. *Exposure to toxins produced by burn pits: Congressional data request and studies.* Memorandum for the assistant secretary of defense for health affairs. Washington, DC: The Joint Staff. March 28, 2011. Enclosure: ASD(HA) Memorandum, 17 Feb 11. Response to ASD(HA) Request for Information.

IOM (Institute of Medicine). 2010. *Gulf War and health: Volume 8: Update of health effects of serving in the Gulf War.* Washington, DC: The National Academies Press.

8

Feasibility and Design Issues for an Epidemiologic Study of Veterans Exposed to Burn Pit Emissions

The charge of this committee is to assist the Department of Veterans Affairs (VA) with design of an epidemiologic study of potential long-term health effects of exposure to burn pit emissions, with specific reference to Joint Base Balad (JBB). The major challenges to the design of such a study are in the areas of exposure assessment and outcome ascertainment. This chapter describes feasibility and design issues associated with such a study. Available datasets on the potential long-term health effects of JBB burn pit emissions are discussed, as are limitations to these data.

Although the chapter focuses on an epidemiologic study of persistent health outcomes in military personnel and veterans associated with exposure to JBB burn pit emissions, the closure of the burn pit in 2009 precludes the collection of further data on concentrations of air pollutants at the time of the operation of the burn pit. Therefore, the committee also considers the possibility of an epidemiologic study of military personnel and veterans exposed to burn pit emissions at any U.S. military base with operating burn pits, not just JBB.

ELEMENTS OF AN EPIDEMIOLOGIC STUDY AND FEASIBILITY ISSUES

The elements of a well-designed epidemiologic study of the potential health effects of an environmental exposure include identification of a relevant study population of adequate size, comprehensive assessment of exposure, careful evaluation of health outcomes, adequate follow-up time, reasonable methods for controlling confounding and minimizing bias, and appropriate statistical analyses (IOM 2000a).

Epidemiologic Study Design Elements

The major types of epidemiologic studies are cohort, case-control, cross-sectional studies (study designs are discussed in detail in IOM 2000a). Most useful to investigating health effects from exposure to burn pit emissions would be a cohort study, such as the epidemiologic studies conducted by the Department of Defense (DoD) at JBB (AFHSC et al. 2010), in which the occurrences of health outcomes among all exposed personnel, ideally with quantitative individual measures of exposure, are compared to control (or unexposed) personnel with similar characteristics. Since the investigation may involve rare health outcomes such as brain cancer, the exposure history of personnel with a disease can be compared to that of non-diseased personnel in a case-control study. Cross-sectional (a snap shot of disease prevalence) and ecologic investigations (analysis of disease risk and exposure at the population level rather than at the individual level) cannot provide strong evidence of a link between burn pit exposure and a specific disease, but they may be useful for detecting possible increases of disease in groups generally assumed to be exposed (such as having ever been deployed to a site with an operating burn pit).

Identification of Study Populations

The selection of appropriate study and control groups is an essential step in the design of a study of military personnel exposed to burn pit emissions. The study sample should be representative of the population of interest and large enough to ensure adequate statistical power. In this case, the population of interest is military personnel who were stationed at JBB during the operation of the burn pit (2003–2009). The committee was provided with information from the Defense Manpower Data Center (DMDC) that indicated the U.S. military population within 10 miles of JBB increased from about 240 in 2003 to about 15,000 in 2007, and then it decreased to about 10,000 in 2009 (Steve Halko, Defense Manpower Data Center, personal communication, August 25, 2010). According to DMDC, the population at JBB might have reached 25,000 when coalition and host-nation forces, civilians, and contractors were included. Many of those stationed at JBB remained on the base for the duration of their tour of duty. With the cooperation of the DMDC, specific information on number of personnel deployed to various locations in or near JBB, the dates and number of their deployments, and other relevant deployment information could be obtained and used to identify an appropriate study cohort.

Military personnel are usually not comparable to the general U.S. population because of a number of factors, such as age and sex distributions, patterns of activity, and behaviors such as smoking, that make the selection of an appropriate control group particularly important and problematic. Although comparisons with the general population are not helpful, comparison within military populations suffer similar issues. For example, deployed military personnel may be healthier in terms of medical history and levels of fitness than nondeployed personnel, a "healthy warrior effect" that may operate even within the military (see Chapter 6). Furthermore, deployed personnel are also exposed to a variety of environmental and personal conditions that the general population and nondeployed personnel may never encounter (IOM 2010). Deployed and nondeployed military personnel may also experience significantly more stressors than the general population. According to a previous IOM report (IOM 2008a) "exposure to combat has been described as one of the most intense stressors that a person can experience; for many people, combat is the most traumatic event of their life." Thus, many researchers have elected to use military personnel deployed to other places as controls when studying military exposures and health effects.

For the study of long-term health outcomes associated with exposure to burn pit emissions, the most appropriate comparison population would be deployed military personnel or veterans who have not been exposed to burn pit emissions either because they were stationed at military bases without burn pits or they were stationed at JBB before or after the burn pits were in operation. Recruiting a control group from this population would reduce the potential for a healthy-warrior effect, as both the exposed and comparison groups of deployed personnel would be similar in terms of their baseline health status.

As an alternative to using an unexposed control group, the committee also suggests a study based on comparisons between subgroups of deployed individuals with different degrees of exposure to burn pit emissions. If exposure can be estimated quantitatively, the wider the range of exposure, the greater the power of the study to detect an association, as described further below.

The committee believes the successful identification of exposed personnel and unexposed personnel deployed to sites without burn pits to be feasible. The DoD has already identified a cohort as evidenced in the report of epidemiologic studies (DoD 2010).

Sample Size

A previous IOM report stated "sufficient samples sizes for each cohort in the study are crucial to ensure adequate statistical power to find differences as well as to reliably identify the lack of differences between groups" (IOM 2000a). Sample-size calculations can be based on "expected magnitude of the difference between the exposed and unexposed groups, the relative sizes of the groups to be compared, and specified levels for type I error (the error of rejecting the null hypothesis when it is true) and type II error (the error of failing to reject the null hypothesis when the alternative hypothesis is true)" (IOM 2000a). Several of the health outcomes potentially related to exposure to burn pit emissions identified in Chapter 6 (asthma, chronic obstructive pulmonary disease [COPD], and cardiovascular disease) are sufficiently common that if large numbers of military personnel or veterans deployed to JBB can be recruited, there should be sufficient power to study the effect of such exposure. Other outcomes of interest, such as various cancers or neurological diseases, may be too uncommon for there to be adequate power to assess whether they are associated with burn pit exposures.

The committee recommends that a pilot feasibility study be conducted to assess whether there will be sufficient power to study specific health outcomes given the currently unknown number of military personnel who might have been exposed to burn pit emissions.

Exposure Assessment

Exposure assessment characterizes the frequency, magnitude, and duration of exposure to an agent of concern in a population. Accurately characterizing exposure is an essential step in conducting a well-designed environmental epidemiologic study. There are a variety of methods for collecting exposure information, but the most desirable is to measure exposures quantitatively at the individual level. Individual exposure measurements can be obtained through personal monitoring data or biomonitoring. However, if individual monitoring data are not available, and they rarely are, individual exposure data may also be estimated from modeling of exposures, self-reported surveys, interviews, job exposure matrices, and environmental monitoring. At JBB, the only environmental monitoring data currently available for the period when the burn pit operated are limited numbers of measured 24-hour average air concentrations (see discussion of environmental monitoring data for JBB in Chapter 4).

Long-term average concentrations are likely to be predicted by dispersion modeling, so that such modeling might be useful in estimating gradients across JBB in long-term average exposures. If further information on the time course of quantities of material burned becomes available (that is, if records are available, or if it can be confirmed that the total amount burned was approximately proportional to the numbers of persons on the base), such information could be incorporated in the dispersion modeling, although there is no guarantee of a linear relationship between emissions and quantities burned. Such estimates of a concentration gradient might be combined with location information for persons on the base (either self-reported or based on job and housing location) to estimate differential exposures to individuals.

Biomonitoring assesses an individual's exposure to environmental agents by measuring the concentrations of the agents in biological samples, usually blood or urine but possibly adipose tissue, hair, or nails. The biomarker can be the external substance itself (for example, lead), or a metabolite of the external substance processed by the body (for example, cotinine, a metabolite of nicotine) and it indicates the absorbed dose or allows an estimate of target-tissue dose for the time of exposure. A chemical with a short half-life in the body might be detected for only a short time after exposure and be indicative of only recent exposure; whereas, chemicals with a long half-life that are not readily metabolized or excreted tend to stay in the body and may be indicative of past and cumulative exposure. For example, serum level of dioxin, which has a long half-life in the body, has been used as a biomarker of exposure to Agent Orange for some Vietnam War veterans (Henriksen et al. 1997). As described in Chapter 4, dioxins have been measured in the air at sampling sites near the JBB burn pit; these environmental measurements might be used as a marker of exposure to burn pit emissions because there is a concentration gradient with distance from the burn pit, taking wind direction into account.

Biomonitoring may be conducted using blood specimens that are available for military personnel through the DoD Serum Repository (see below). More research is needed to identify useful, meaningful, reliable, and implementable biomarkers and methods to measure exposure to dioxins and other chemicals in burn pit emissions. The value of serum dioxin as a biomarker of exposure for residents living near municipal incinerators has been studied in Spain and Taiwan, but serum levels were not well correlated with either measured atmospheric concentrations of dioxins or distance from the incinerators (Gonzalez et al. 2000; Huang et al. 2007). An initial attempt to use serum dioxin measurements in JBB personnel did not provide useful information (Taylor et al. 2008), and the predicted increments in serum concentrations based on an estimate of air exposures are low compared with background serum concentrations in the general U.S. population. However, the serum dioxin measurements were severely compromised by high detection limits. The committee suggests that a paper exercise be conducted to estimate whether the expected small increment in serum concentrations, with the expected congener signature (based on the air measurements and the physiologically based pharmacokinetic model for dioxin), would be detectable if all 2,3,7,8 congeners could be accurately measured. If so, it would also be useful to determine if special efforts would now allow adequate detection of all the congeners in the small serum specimens available.

Personal monitors make it possible to directly measure exposures for a study population at the interface between the exposure medium and the human receptor (e.g., the breathing zone). In conjunction with records of the individuals' activities, locations where the exposure to the highest concentrations might occur as well as the nature of emission sources can often be inferred. Personal monitoring methods are subject to various constraints, principally the availability of sensitive, cost-effective, easy to operate, sample collection equipment and analytic methods that provide sufficient time resolution and are free from interferences. In the case of JBB burn pit exposures, personal monitoring can no longer be employed because the burn pit is no longer operational. However, burn pits are still in operation at other military bases and personal monitoring at those bases, particularly for potentially highly exposed populations such as the pit operators, could be conducted.

Environmental monitoring is an indirect method of exposure assessment. Estimates of exposure may be made by combining measurements of pollutant concentrations at fixed sites with information on rates of contact with the medium of interest (e.g., air, water, soil) and recorded time-activity information. Modeling of air concentrations of particular chemicals at specified sites combined with workers' job histories can be used to estimate the cumulative dose for each worker. At JBB, air pollutant concentrations at the three sampling locations, or based on dispersion modeling as described above, could be combined with information from personnel records on job type (activity), job location (that is, distance from the burn pit), and time stationed at JBB (duration of exposure) to assign burn pit exposures to individuals for an epidemiologic study.

Further environmental monitoring might be possible to evaluate gradients of exposure across the base; for example, it may be possible to evaluate the gradient in long-term average exposure to the burn pit by measurement of soil concentrations of PCDDs/Fs combined with the air dispersion modeling described above. The rate at which PCDDs/Fs deposit to soil depends on air concentration at ground level and the size of ambient particles on which the PCDDs/Fs condense (which particles are a complex mixture of the particles emitted by the burn pit and ambient particles from other sources). Once mixed in soil, PCDDs/Fs tend to be stable, but the mixing rate into the subsurface is location dependent. A pilot study to measure dioxin in soil across the base is likely the only approach that would definitively demonstrate whether gradients can be adequately established.

Compared with personal monitoring, use of environmental concentrations loses information because personnel at a given site may experience varied levels of exposure based on their activities, personal characteristics, and day-to-day differences in ambient pollutant concentrations; use of environmental concentrations assumes personnel are exposed to the average exposure at each site. Any approach that blurs the distinction between individual exposures while maintaining the collection of individual health outcomes will reduce the estimated variation in exposure and lead to exposure misclassification of some individuals. That biases the study, increasing the chance that any association between exposure and health outcome will not be detected.

Other approaches, such as self-report questionnaires and review of military-activity records by themselves, that is, those not linked to environmental monitoring, are unlikely to yield as accurate an assessment of exposure because of recall bias and greater potential for exposure misclassification. The usefulness of such historical information for exposure assessment might be enhanced with linkage to maps of JBB that display the location of the burn pit and the three sampling locations (mortar pit, guardhouse/transportation field, and CASF/H-6 housing). Unfortunately, the environmental monitoring conducted at JBB was done on an insufficient number of days (that is, sampling was only done on 53 days in 2007 and 2009 combined) to provide reliable estimates of long-term average exposures to burn pit emissions.

Given the limitations described, in the opinion of the committee, the sampling data described in Chapter 4 are sufficient only to allow ordinal categorization of exposure (that is, low, moderate, high) in combination with questionnaires and time-activity information. Improved quantitative estimates might be obtained if air dispersion and deposition modeling are consistent with gradients in soil dioxin measurements. Considering the lack of currently available exposure information, additional environmental monitoring at other sites with operating burn pits will provide greater context for future studies as would biomarkers if they can be reliably correlated with environmental exposure. Feasibility will be greatly affected by the availability, reliability, and utility of the data sets and other information resources.

Evaluation of Health Outcomes

In addition to accurate and comprehensive exposure information, accurate ascertainment of the health outcome(s) of interest is necessary for the conduct of an epidemiologic study of the potential health effects of an environmental agent. A variety of methods for ascertaining health outcomes is available, including review of death certificates, medical records, and data from clinical examinations; linkage to disease registries; and use of self-reported outcomes from surveys or interviews.

Medical records can be a useful source of outcome data because health information in these records is usually recorded by trained health-care providers, although recording errors or misdiagnoses may occur. Death certificates are comprehensive in coverage but do not capture nonfatal adverse health outcomes, contain limited data, and coding errors are common. Like self-reported exposure information, self-reported outcome information may be biased because subjects do not always recall information accurately. It is therefore important to verify self-reported information through physician diagnoses, death certificates, or disease registries when feasible.

Complete and accurate assessments of health outcomes for exposed and control groups are important to evaluate whether an increase in the number of cases of a particular disease is related to exposure to burn pit emissions. That information can then be compared with the background rate of the disease in the control population to determine whether there is an "excess" of cases in the exposed population. Outcome assessments should be designed to minimize bias in ascertainment and have an adequate follow-up period to allow for the observation of outcomes with long latency periods, such as cancer.

Numerous studies have demonstrated that military personnel and their health status can be successfully identified for prospective and retrospective epidemiologic studies, for example, the Millennium Cohort Study (MCS) (see discussion of this study later in the chapter) and the recent epidemiologic studies conducted by the DoD (AFHSC et al. 2010). Given the many outcome assessment resources available to the DoD (described below), the committee believes the long-term follow-up of personnel exposed to burn pits to be a feasible task.

Minimizing Bias

An observed association between burn pit exposure and a health outcome could be confounded by factors that are related to both the likelihood of exposure to burn pit emissions and the outcome. For example, military personnel who serve in roles that pose an increased risk of burn pit exposure may be more likely to be smokers, and smoking is associated with increased risk of respiratory disease, cardiovascular disease, and some cancers. The influence of confounders on the relationship between burn pit exposure and health outcomes can be reduced by careful study design and data-analysis schemes. Multivariable data-analysis techniques, such as stratified analysis and regression modeling, can effectively control for confounding by adjusting for multiple factors simultaneously (for example, age, smoking, and sex). However, the effectiveness of such methods depends in part on ascertaining potential confounding factors with accuracy and precision. Furthermore, controlling for confounding factors is of greater importance in situations where the magnitude of the association, in this case burn pit emissions and health outcomes, is modest relative to the relationship between the confounder and either burn pit exposure or the health outcome. For example, smoking is strongly associated with respiratory disease so it may be more difficult to attribute an increase in respiratory disease among exposed military personnel who smoke to burn pit emissions compared with nonsmokers.

Effect modification should be minimized in the design of an epidemiologic study. For example, sex would be an effect modifier if there were a substantial difference in the magnitude of the association between burn pit exposure and a health outcome for men and women. When such interactions are detected, effects must be estimated and reported separately by subgroup (that is, stratified) rather than adjusted for in the whole population, as is appropriate for a confounder; such stratified analyses would require an even larger total study size to have adequate power to detect effects. Obtaining large numbers of study subjects poses a challenge for any epidemiologic study even in the absence of interactions.

While controlling for the influence of all factors is not practical, the DoD must attempt to minimize bias by controlling for the most influential confounding exposures. The committee suggests that control for confounding or effect modification should be considered and integrated into the study design. Of particular interest are behavioral habits (smoking, activity level), personal characteristics (age, race, sex), and other environmental or occupational exposures (for example, occupations with exposure to toxicants such as jet fuel). The feasibility of identifying and controlling for specific confounders should be addressed in a pilot study; the resulting data can be used to allow modification of the study design or adjustment of the analyses.

AVAILABLE DATASETS AND RESOURCES

Available datasets that may provide relevant information on either exposures to burn pit emissions or long-term health outcomes associated with such exposures are discussed below. Ongoing and planned studies of active-duty military personnel or veterans that can be used to assess long-term health outcomes associated with burn pit emissions are also described. The use of existing data or assessment efforts for future studies is helpful in reducing costs and enhancing study feasibility.

Exposure Assessment Information Sources

As described in Chapter 4, environmental sampling in the vicinity of the JBB burn pit was conducted from 2006 through 2009 by the U.S. Army Center for Health Promotion and Preventive Medicine (CHPPM, now the U.S. Army Public Health Command) (Taylor et al. 2008; CHPPM and AFIOH 2009; USAPHC 2010). No individual exposure data were collected at JBB. The burn pit was closed in 2009, and thus there are no further opportunities to collect environmental or personal monitoring data to assess exposure to burn pit emissions or other air pollution at JBB, except perhaps to measure concentrations of dioxin in soil and correlate those with air dispersion and deposition modeling.

However, other sources of exposure data do exist. Launched in 2001, the MCS is a coordinated, systematic 21-year effort to study the potential long-term health effects associated with deployment-related exposures (http://www.millenniumcohort.org/index.php, accessed November 11, 2010). The cohort consists of 152,000 consenting military personnel drawn from all branches of military service, including the Coast Guard (Ryan 2007), and representing approximately 11.3% of the 2.2 million men and women in service as of October 1, 2000. Recently, the MCS has focused on burn pit emissions at JBB and has been assigning exposures to military personnel stationed there by creating geographic buffer areas of varying radii (2, 3, 5, or 10 miles) around the burn pit. Those who resided within a specific buffer zone are considered to be exposed to burn pit emissions compared with those not residing within the buffer (DoD 2010). Duration of time spent at the JBB within the buffer zone can also be used to assess exposure for a study of long-term health effects. Estimates of exposure to JBB burn pit emissions could be enhanced by the inclusion of data on wind speed and direction, which are available for JBB. Land-use information could also be incorporated into the exposure assessment to provide a more complete characterization of possible sources of particulate matter present on the base.

The MCS also collects self-reported exposure data in its surveys of study participants. Beginning in 2010, the MCS survey has included a question on "exposure to burning trash." Certain potentially highly exposed subgroups, such as personnel who regularly visited or who operated the burn pit, could be identified by further queries on exposure. Two DoD questionnaires, the Post-Deployment Health Assessment (PDHA) and the Post-Deployment Health Reassessment (PDHRA), are administered to all deployed military personnel upon their return to the United States and again 3–6 months later (available at http://www.pdhealth.mil/dcs/post_deploy.asp). Those questionnaires include a number of items about health outcomes and exposure, including "Are you worried about your health because you were exposed to 'smoke from burning trash or feces'?" The Armed Forces Health Surveillance Center (AFHSC) collects the PDHA and PDHRA data (see section on Evaluation of Health Outcomes).

Another factor that affects the exposure of military personnel at a base with a burn pit is the type of housing available. While the committee has not done a systematic review of the housing at bases in Iraq and Afghanistan, it is likely that much of the housing would allow penetration of fine particles into the living spaces. Thus, military personnel would be exposed during off-duty time spent in their quarters as well as during duty hours. Further information on housing available on bases and indoor air monitoring would also help characterize potential exposures to burn pit emissions.

Biomonitoring

The DoD Serum Repository collects and stores specimens from military personnel; it currently houses over 50 million specimens linked to individual demographic data. The mission of the repository is "to receive and store remaining serum specimens from HIV testing programs within the DoD, and to receive and store serum specimens related to operational deployments worldwide" (available at http://afhsc.army.mil/dodsr). The AFHSC states that the repository can be used in epidemiologic studies and has been used historically as a screening tool to identify widespread infection.

A potential approach to exposure assessment for a future study of persistent effects of burn pit emissions is to measure one or more environmental agents or their metabolites (biomarkers of exposure) in samples of serum. One could also measure biomarkers of effect such as C-reactive protein or fibrinogen, known biomarkers of cardiovascular disease risk, assuming the study design allows for the biomarker of effect to be correlated with exposure.

The MCS can already be linked to the DoD Serum Repository. The repository has approximately four samples stored per individual, although for many individuals pre- and post-deployment samples are not available. The MCS is exploring the feasibility of using the Serum Repository samples for biomonitoring of military personnel deployed at the JBB (Smith et al. 2009).

Evaluation of Health Outcomes

Medical Surveillance of Active-Duty Military Personnel

The AFHSC is the central epidemiologic resource for the DoD; its main functions are to "analyze, interpret, and disseminate information regarding the status, trends, and determinants of the health and fitness of U.S. military (and military-associated) populations, and to identify and evaluate obstacles to medical readiness" (http://afhsc.army.mil/viewDocument?file=AFHSC_Brochure/AFHSC_Brochure.pdf). The AFHSC maintains a number of surveillance databases that are used to track the health of military personnel, including the PDHA and PDHRA described earlier. One such database is the Defense Medical Surveillance System (DMSS) that contains up-to-date and historical data on diseases and medical events (for example, hospitalizations, reportable diseases, and acute respiratory diseases, among others), longitudinal data on personnel and deployments, and the Standard Inpatient Data and Standard Ambulatory Data Records (available at http://afhsc.army.mil/dmss). AFHSC uses the DMSS to provide a link between the DoD Serum Repository and other databases. The DMSS records could be used to study health effects in those exposed to JBB burn pit emissions, although given the closure of the burn pit in late 2009, such a study could only be retrospective.

The DoD maintains a number of registries, including the Automated Central Tumor Registry and the DoD Medical Mortality Registry. The former registry is the DoD's central registry for cancer and may be linked to other databases so cancer data can be tracked. The DoD Medical Mortality Registry, administered by the Armed Forces Medical Examiner System, collects complete medical and related information on every military active-duty death for surveillance and prevention purposes.

Post-Discharge Surveillance

In addition to civilian disease and death registries, several databases supported by the VA that follow veterans after they have left the military would be useful for an epidemiologic study and assist in long-term follow-up.

The VA Beneficiary Identification and Record Locator Subsystem (BIRLS) is an automated system for identifying veterans and their beneficiaries who have received compensation, pension, education, or other VA benefits. The BIRLS database contains the BIRLS Death File. Veteran disability benefit claims for health outcomes potentially related to burn pit exposure (for example, respiratory diseases) can be obtained for veterans who have been identified as exposed to burn pits.

The VA Patient Treatment File (PTF) contains information on inpatient records for each discharge from a VA hospital facility since 1970 (updated biweekly). It can be used as a sampling frame to identify potential subjects for case-control studies. The PTF can also be used to assess health care utilization or morbidity for selected cohorts of veterans. The VA Outpatient Clinic File contains records from visits to outpatient VA clinics.

The VA maintains several health registries. The Cancer Registry is compiled from cancer registries maintained at each VA facility and contains records on all veterans receiving a diagnosis of cancer within the VA medical system. The VA also maintains registries of veterans diagnosed within the VA system with other conditions, including amyotrophic lateral sclerosis and multiple sclerosis.

The VA also supports a program called War-Related Illness and Injury Study Centers (WRIISC). The goal of WRIISC research is to better understand and improve the health of combat veterans. This research investigates topics of concern for combat veterans and their families, health care providers, and policy makers. WRIISC research interests include: environmental exposures and post deployment health and the long-term health effects of war.

Cohort Studies

The MCS described earlier has already been used to investigate the health outcomes among OIF/OEF personnel at bases with burn pits, but the followup in these investigations was limited (Smith et al. 2009). Use of the MCS is limited by its participation rate; initially enrollment was only 35%, although the follow-up rate is approximately 70–80%. Thus, the retained participants may represent a biased sample of military personnel, potentially different from military personnel who chose not to participate. The MCS is planning to enroll an additional 50–60,000 participants, bringing the total study population to 210,000. These new enrollees will be asked to donate biological samples for biomarker assays (Smith et al. 2009).

Outcome assessment is based on self-reports, although the MCS can also access DoD inpatient, pharmaceutical, and vaccination data through the Defense Medical Surveillance System. Unfortunately, those data are only available for active-duty military personnel. At present, the MCS is not able to access health outcome data in VA databases.

The VA is conducting The National Health Study for a New Generation of U.S. Veterans, a 10-year longitudinal study of 60,000 OIF and OEF veterans. A pilot study with 3,000 participants was completed in spring 2009; the response rate among contacted eligible veterans was about 30%. The full-scale study began with a questionnaire in 2009. Exposure and outcome data are entirely self-reported and based on a series of question about types of health care, diagnoses, and a variety of exposures, including a question about exposure to burning trash or feces, oil fires, and smoke. The survey also collects information on branch of the military, component, and job title, and asks about where the respondents were stationed since 2001 and the total number of times deployed. The validity and reliability of the questionnaire is being studied.

PROPOSED APPROACHES TO THE STUDY OF HEALTH OUTCOMES FROM EXPOSURE TO BURN PIT EMISSIONS

The committee recommends a cohort study of the long-term health effects (evaluated prospectively) from retrospective estimates of exposure to burn pit emissions in military personnel deployed at the JBB. To determine the incidence of chronic diseases or cancers with long latency, individuals must be followed for many years. Ideally, the observation period for health effects begins retrospectively, at first deployment to JBB, and continues after active duty is completed. First, however, the committee strongly recommends that pilot studies be conducted to address issues of statistical power and develop design features for specific health outcomes. It is important to note that once a prospective cohort infrastructure has been established, multiple health outcomes can be studied in the cohort over time. Intermediate outcomes on the pathway to the development of chronic diseases can also be studied in a serial manner. For example, serial spirometry can be used to detect excessive rates of decline in lung function before a diagnosis of COPD is made and serial measurement of intima-media thickness in carotid arteries can be used to detect early atherosclerosis.

Retrospective estimation of exposure will prove particularly difficult. To characterize exposures to the complex mixture of burn pit emissions while accounting for other environmental and occupational hazards including air pollutants from other sources, the committee recommends a tiered approach. Since direct quantitative exposure measurements are not possible, the committee sought to outline general study designs that could easily be informed using available data previously described. The three tiers of the recommended study are characterized by the decreasing specificity of exposure and would answer different research questions, as follows:

- **Tier 1:** *Did proximity to burn pit operations at JBB increase the risk of adverse health outcomes?*

    Ordinal estimates of individual-level exposure to JBB burn pit emissions can be determined based on dates of deployment, duties on base, and location of housing relative to the burn pit, taking account of wind-dispersion models. The exposure effect can be assessed by comparing subgroups with more and less exposure among all individuals with potential exposure, that is, those stationed at the JBB during the period of full burn pit operation (2003–2007). The use of soil dioxin concentrations at various locations at JBB should be considered as a potential marker of exposure to burn pit emissions.

- **Tier 2:** *Did installation of incinerators at JBB reduce incidence of disease or intermediate outcomes (for example, emphysema or rate of lung function decline)?*

    Assess exposure (yes/no) to JBB burn pit by date of initial deployment. This approach considers the installation of incinerators to replace the burn pit as an intervention. Comparisons can be made between post-deployment chronic health outcomes among those deployed to JBB before the burn pits were shut down and those deployed there afterwards. A caveat here is that the installation of incinerators at JBB occurred in phases. In July 2007, two incinerators were put into operation at JBB, and in April 2008 a third incinerator began operation. By October 2009, burn pit operations at JBB ceased when the fourth incinerator began operating, resulting in 100% solid waste disposal via incineration or off-site recycling. Thus, it may be necessary to study three different times defined by the beginning and end of the incinerator installation period: before July 1, 2007; July 1, 2007, to October 1, 2009; and after October 1, 2009.

- **Tier 3:** *Did deployment at JBB during full burn pit operation increase the risk of adverse health outcomes compared with deployment elsewhere in Iraq or Afghanistan or with no deployment?*

    Assess exposure (yes/no) to the total JBB environment, recognizing that the burn pit emissions occurred in the presence of PM and other air pollutants from other sources. This broad definition of exposure can be assessed by comparing the health experience of military personnel deployed at the JBB during the period of burn pit operation to that of military personnel deployed to Iraq and Afghanistan at locations without a burn pit or that of military personnel not deployed to the Middle Eastern theatre during the same time. This approach was used by AFHSC and MCS to conduct short-term health studies described in Chapter 6 (AFHSC et al. 2010). Although there are limitations to this approach, it may be possible to find an appropriate unexposed comparison group—preferably another deployed population unexposed to burn pits but exposed to PM and other chemicals identified at JBB from other sources. The recommendation for a nondeployed comparison group is based on the committee's judgment that pollution in the region from sources other than burn pits may pose greater health risks than burn pit emissions.

Elements of a three-tiered prospective cohort study of active-duty military personnel and veterans designed to assess potential chronic health effects related to burn pit emissions are presented in Table 8-1.

TABLE 8-1 Proposed Nested Prospective Cohort Studies of Long-Term Health Effects

Tier	Question	Population	Exposure
1	Did proximity to burn pit at JBB increase risk of adverse outcomes?	Deployed to JBB during burn pit operation (2003–2007)	Estimated individual level ordinal exposure to burn pit emissions based on JBB job, barracks location, and duration of JBB deployment. Exposure: low, medium, high
2	Did installation of incinerators at JBB reduce disease incidence or rate of change in intermediate outcomes?	Deployed to JBB before, during and after burn pit operation (2003–2009)	Deployed to JBB before, during, or after incinerator installation. Exposure: yes or no
3	Did deployment at JBB during full burn pit operation increase risk of adverse outcomes compared to deployment elsewhere in Iraq or Afghanistan or compared to nondeployed?	Military service during 2003–2007 (deployed and nondeployed)	Deployed to JBB with burn pit, deployed to Iraq or Afghanistan sites without burn pits, or nondeployed. Exposure: yes or no

FEASIBILITY ISSUES

The committee acknowledges that greater specificity in study design is accompanied by greater limitations on data collection and decreased feasibility in conducting the study. Feasibility issues for an epidemiologic study to determine the long-term consequences of exposure to burn pits in Iraq and Afghanistan include the inability to obtain direct, individual exposure measurements for personnel at JBB as the burn pit has been closed since 2009. However, it might be feasible to obtain job title, duties, and base location in addition to the location of barracks to estimate individual exposure. From previous studies, the most simplistic determination of exposure to burn pits, that is deployment to a site operating a burn pit (yes or no), is feasible as described in Tier 3 (DoD 2010). Furthermore, as noted earlier, there are other military bases with operating burn pits, particularly in Afghanistan where personal monitoring could be conducted.

The main determinant of study feasibility is access to complete, accurate, reliable data pertaining to deployment, demographic and personal characteristics, and health outcomes from the DoD, the VA, and civilian sources. The committee recommends a pilot study be completed to examine feasibility including data availability, power, exposure assessment, confounding, and assessment of health outcomes.

CONCLUSIONS AND RECOMMENDATIONS

As outlined by the committee's statement of task, several important aspects of epidemiologic study design and their feasibility have been discussed. Considering the feasibility issues, the committee has developed the following recommendations for a potential study of long-term health effects associated with exposure to burn pits:

- A cohort study of veterans and active duty military should be considered to assess potential long-term health effects related to burn pit emissions in the context of the other ambient exposures at the JBB. This type of study, while complex, is not unique in a military setting (for example, standard methods exist for the U.S. Air Force Ranch Hand Study that examined health effects of Agent Orange).
- An independent oversight committee composed of military and external experts in air pollution, analytical chemistry, exposure assessment, epidemiology, toxicology, biostatistics, and occupational and environmental medicine should be established to provide guidance and to review specific objectives, study designs, protocols, and results from the burn pit emissions research programs that are developed. Such a committee

could provide an essential peer-review function to lend greater scientific credibility to the investigations. An example is the advisory committee that was established to oversee the conduct of the Ranch Hand Study (IOM 2006).
- A pilot study should be conducted to ensure adequate statistical power, ability to adjust for potential confounders, to identify data availability and limitations, and develop testable research questions and specific objectives. The objectives should be used to motivate essential study design features. Examples of these features include: subject eligibility criteria, size and demographic characteristics of the cohort, length of follow-up required, health outcomes to be studied, critical time periods of exposure, and potential confounding and modifying factors that would need to be measured. Careful consideration should be given to defining sensitive and useful exposure measures.
- Assessment of health outcomes is best done collaboratively using the clinical informatics systems of the DoD and the VA, in addition to the non-military methods of follow-up (for example, National Death Index, state cancer registries) that can be used to identify the incidence and prevalence of health effects over time. Integration of current programs, such as the MCS, would increase feasibility and ease of study initiation. Multiple health assessments in the form of questionnaires and specific medical assessments could be administered periodically to better address intermediate and non-fatal health outcomes.
- An exposure assessment for better source attribution and identification of chemicals associated with waste burning and other pollution sources at JBB should be conducted prior to beginning a new epidemiologic study to help the VA determine those health outcomes most likely to be associated with burn pit exposures. The committee's analysis of available data from the environmental monitoring conducted at JBB suggests that exposure to PM emitted from sources such as diesel and jet engines, upwind Iraqi urban areas, and soil, may be of greater concern than exposure to burn pit emissions.
- Exposure assessment should include detailed deployment information including distance and direction individuals lived and worked from the JBB burn pit, duration of deployment, and job duties. Multiple methods of estimating exposure have been discussed; however, the most applicable method should be defined by the study questions, data availability and limitations, and study design. Study of troops currently deployed at bases with operating burn pits, in addition to JBB, would allow for prospective exposure assessment of those troops and provide information useful to interpretation of results from JBB.

In conclusion, a study of health effects resulting from exposure to burn pits is feasible but its ability to produce useful and actionable results depends on a well thought-out design, thorough exposure assessment and careful follow-up. The IOM and NRC have recommended several methodologies for investigating and monitoring the health of military service members to the VA and the DoD over the years that, in addition to this report, can provide further guidance on study design and feasibility (IOM 1999, 2000b, 2008b; NRC 2000a,b).

REFERENCES

AFHSC (U.S. Armed Forces Health Surveillance Center), the Naval Health Research Center, and the U.S. Army Public Health Command. 2010. *Epidemiological studies of health outcomes among troops deployed to burn pit sites*. Silver Spring, MD: Defense Technical Information Center.

CHPPM (U.S. Army Center for Health Promotion and Preventive Medicine) and AFIOH (U.S. Air Force Institute for Operational Health). 2009. *Addendum 2. Screening health risk assessment burn pit exposures Balad Air Base, Iraq, May 2008*. USACHPPM Report No. 47-MA-08PV-08/AFIOH Report No. IOH-RS-BR-TR-2008-0001. Aberdeen Proving Ground, MD: U.S. Army Center for Health Promotion and Preventive Medicine. August.

DoD (U.S. Department of Defense). 2010. *Report to Congress on the use of open-air burn pits by the United States armed forces*. Washington, DC: U.S. Department of Defense.

Gonzalez, C. A., M. Kogevinas, E. Gadea, A. Huici, A. Bosch, M. J. Bleda, and O. Papke. 2000. Biomonitoring study of people living near or working at a municipal solid-waste incinerator before and after two years of operation. *Archives of Environmental Health* 55(4):259-267.

Henriksen, G. L., N. S. Ketchum, J. E. Michalek, and J. A. Swaby. 1997. Serum dioxin and diabetes mellitus in veterans of Operation Ranch Hand. *Epidemiology* 8(3):252-258.

Huang, H. Y., T. Y. Jeng, Y. C. Lin, Y. C. Ma, C. P. Kuo, and F. C. Sung. 2007. Serum dioxin levels in residents living in the vicinity of municipal waste incinerators in Taiwan. *Inhalation Toxicology* 19(5):399-403.

IOM (Institute of Medicine). 1999. *Gulf War veterans: Measuring health*. Washington, DC: National Academy Press.

IOM. 2000a. *Gulf war and health: Volume 1. Depleted uranium, pyridostigmine bromide, sarin, vaccines*. Washington, DC: National Academy Press.

IOM. 2000b. *Protecting those who serve: Strategies to protect the health of deployed U.S. forces*. Washington, DC: National Academy Press.

IOM. 2006. *Gulf War and health: Volume 4. Health effects of serving in the Gulf War*. Washington, DC: The National Academies Press.

IOM. 2008a. *Gulf War and health: Volume 6. Physiologic, psychologic, and psychosocial effects of deployment-related stress*. Washington, DC: The National Academies Press.

IOM. 2008b. *Epidemiologic studies of veterans exposed to depleted uranium: Feasibility and design issues*. Washington, DC: The National Academies Press.

IOM. 2010. *Gulf War and health: Volume 8. Update of health effects of serving in the Gulf War*. Washington, DC: The National Academies Press.

NRC (National Research Council). 2000a. *Strategies to protect the health of deployed U.S. forces: Analytical framework for assessing risks*. Washington, DC: National Academy Press.

NRC. 2000b. *Strategies to protect the health of deployed U.S. forces: Detecting, characterizing, and documenting exposures*. Washington, DC: National Academy Press.

NRC. 2010. *Review of the Department of Defense Enhanced Particulate Matter Surveillance Program report*. Washington, DC: The National Academies Press.

Ryan, Margaret A. K., T. C. Smith, B. Smith, P. Amoroso, E. J. Boyko, G. C. Gray, G. D. Gackstetter, J. R. Riddle, T. S. Wells, G. Gumbs, T. E. Corbeil, and T. I. Hooper. 2007. Millenium cohort: enrollment begins a 21-year contribution to understanding the impact of military service. *Journal of Clinical Epidemiology* 60:181-191.

Smith, B., C. A. Wong, T. C. Smith, E. J. Boyko, G. D. Gackstetter, M. A. K. Ryan, and Team Millennium Cohort Study. 2009. Newly reported respiratory symptoms and conditions among military personnel deployed to Iraq and Afghanistan: A prospective population-based study. *American Journal of Epidemiology* 170(11):1433-1442.

Taylor, G., V. Rush, A. Deck, and J. A Vietas. 2008. *Screening health risk assessment burn pit exposures Balad Air Base, Iraq and addendum report*. IOH-RS-BR-TR-2008-0001/USACHPPM 47-MA-08PV-08. Brooks City-Base, TX: Air Force Institute for Operational Health and U.S .Army Center for Health Promotion and Preventative Medicine.

USAPHC (U.S. Army Public Health Command). 2010. *Screening health risk assessments, Joint Base Balad, Iraq, 11 May–19 June 2009*. Aberdeen Proving Ground, MD: U.S. Army Center for Health Promotion and Preventive Medicine. July.

Appendix A

Committee Biographical Sketches

David J. Tollerud, M.D., M.P.H. (*Chair*), is a professor and chair of the Department of Environmental and Occupational Health Sciences at the School of Public Health and Information Sciences at the University of Louisville in Louisville, Kentucky. He has extensive clinical training, with specialty board certifications in internal medicine, pulmonary and critical care medicine, and occupational medicine. Dr. Tollerud has research expertise in environmental and occupational health, epidemiology, and immunology, and consulting experience in the areas of occupational and environmental respiratory disease, medical surveillance, and workplace injury prevention programs. He served as the chair of the Committee on the Disposition of the Air Force Health Study, and he has served on the IOM's Board on Population Health and Public Health Practice and on a number of IOM committees since 1992.

John R. Balmes, M.D., is a professor of medicine at the University of California, San Francisco, and chief of the Division of Occupational and Environmental Medicine at San Francisco General Hospital. He is also a professor of environmental health sciences at the University of California, Berkeley, and director of the Northern California Center for Occupational and Environmental Health. Dr. Balmes studies the respiratory health effects of various air pollutants. He has a particular interest in occupational respiratory disease. He has investigated the acute effects of inhalation exposures to ambient air pollutants in his human exposure laboratory at San Francisco General Hospital and the chronic effects of such exposures in epidemiological studies with collaborators at the University of California, San Francisco and the University of California, Berkeley. Dr. Balmes also is investigating genetic determinants of responses to air pollutants. He has led research, funded by the U.S. Centers for Disease Control and Prevention, to assist in the development of a national program to link environmental hazards with health outcome data to improve the tracking of diseases potentially related to environmental exposures. He is also the physician member of the California Air Resources Board. He served on the NRC's Committee for the Review of the Army's Enhanced Particulate Matter Surveillance Project Report. Dr. Balmes received his M.D. from the Mount Sinai School of Medicine.

Aruni Bhatnagar, Ph.D., is a Professor of Medicine and Distinguished University Scholar at the University of Louisville, Kentucky. His research interests include cardiovascular toxicity of environmental aldehydes, including pollutants such as gasoline vapor, car exhaust, smoke, and smog, among others. Some of Dr. Bhatnagar's research tests the relation between exposure to aldehydes and myocardial dysfunction, and exposure to aldehydes as a risk factor for developing heart disease. He received his Ph.D. in chemistry from the University of Kanpur, India. He obtained his post-doctoral training at the University of Texas Medical Branch, Galveston, Texas.

Edmund A. C. Crouch, Ph.D., is a senior scientist with Cambridge Environmental Inc. in Cambridge, Massachusetts. He has published widely in the areas of risk assessment, and presentation and analysis of uncertainties. He has co-authored a major text in risk assessment, *Risk/Benefit Analysis*. Dr. Crouch serves as an expert advisor to various local and national agencies concerned with public health and the environment, and has served on a number of National Academy of Sciences committees, including the Committee on Health Effects of Waste Incineration. Dr. Crouch received his Ph.D. in High Energy Physics from Cambridge University.

Francesca Dominici, Ph.D., is a professor of biostatistics in the Harvard School of Public Health at Harvard University. Dr. Dominici's research has focused on the interface between the methodological development of hierarchical models and their applications to multi-level data. She has extensive experience on the development of statistical methods and their applications to clinical trials, toxicology, biology, and environmental epidemiology. Her main research interest is in the development of statistical models and the conduct of epidemiological studies to estimate the health effects of air pollution. She has served on a number of National Academies' committees including the Committee on Gulf War and Health: Review of the Medical Literature Relative to Gulf War Veterans' Health; the Committee to Assess Potential Health Effects from Exposures to PAVE PAWS Low-Level Phased-array Radiofrequency Energy; and the Committee on The Utility of Proximity-Based Herbicide Exposure Assessment in Epidemiologic Studies of Vietnam Veterans. Dr. Dominici received her Ph.D. in statistics at the University of Padua, Italy.

Ellen A. Eisen, Sc.D., is an adjunct professor of environmental health sciences and epidemiology at the School of Public Health at the University of California, Berkeley, and environmental health at Harvard School of Public Health. Her research focuses on methods for modeling exposure-response in occupational cohort studies. She has studied respiratory, cancer and cardiovascular outcomes in relation to a wide variety of occupational exposures, including silica, cotton dust, endotoxin, welding fumes, and metalworking fluids. Her early studies of longitudinal decline in pulmonary function identified high test variability of FEV_1 as a biomarker of impaired lung function and the standard exclusion of nonreproducible tests as a source of selection bias. She has published many papers on cancer incidence and mortality in relation to oil-based metalworking fluids in autoworkers. Dr. Eisen has served on numerous NRC and IOM committees, including the Committee on Asbestos: Selected Health Effects and the Committee on the Health Effects of Mustard Gas and Lewisite. Dr. Eisen earned her Sc.D. in biostatistics and occupational health from the Harvard School of Public Health.

Mary A. Fox, Ph.D., M.P.H., is an assistant professor in the Department of Health Policy and Management at the Johns Hopkins Bloomberg School of Public Health. Her research is focused on developing cumulative risk assessment to inform public-health decision making. Dr. Fox has applied cumulative-risk methods in numerous community health assessments. Her current research is directed at national-level decision-making and includes the relationship between exposure to a mixture of nephrotoxic metals and renal function and applications of risk assessment to policy evaluation. Dr. Fox earned her M.P.H. from the University of Rochester School of Medicine and Dentistry and her Ph.D. from the Johns Hopkins Bloomberg School of Public Health.

Mark W. Frampton, M.D., is a professor of medicine and environmental medicine at the University of Rochester School of Medicine and Dentistry. He is also the medical director of the Pulmonary Function Laboratory at the University of Rochester Medical Center. Dr. Frampton is board certified in Internal Medicine and Pulmonary Diseases, and his research focuses on human clinical studies of the health effects of gaseous and particulate air pollution. He has served as a member on the external scientific advisory committees of the Southern California and Harvard Particulate Matter Centers, and as a consultant for the California Air Resources Board. Dr. Frampton received his M.D. from the New York University School of Medicine.

Petros Koutrakis, Ph.D., is a professor at the Harvard School of Public Health, Head of the Exposure, Epidemiology and Risk Program and the Director of the EPA/Harvard University Center for Ambient Particle Health Effects. His research activities focus on the development of human exposure measurement techniques and the investigation of sources, transport, and the fate of air pollutants. In collaboration with his colleagues in the Environmental Chemistry Laboratory, he has developed ambient particle concentrators and high volume samplers that can be used to conduct human and animal inhalation studies. He has also developed a personal ozone monitor, a continuous fine particle measurement technique and several other sampling methods for a variety of gaseous and particulate air pollutants. These novel techniques have been used extensively by air pollution scientists and human exposure assessors in United States and worldwide. Dr. Koutrakis has conducted a number of comprehensive air pollution studies in the United States, Canada, Spain, Chile, Kuwait, Cyprus, and Greece that investigate the extent of human exposures to gaseous and particulate air pollutants. Other research interests include the assessment of particulate matter exposures and their effects on the cardiac and pulmonary health. Dr. Koutrakis is a member of national and international committees and the past Technical Editor-in-Chief of the *Journal of the Air & Waste Management Association*. He has published more than 170 peer reviewed papers in the areas of air quality, exposure, and health effects assessment and instrumentation. Dr. Koutrakis received his Ph.D. in Environmental Chemistry from the University of Paris in 1984.

Jacob McDonald, Ph.D., is a scientist and director of the Chemistry and Inhalation Exposure Program at Lovelace Respiratory Research Institute. He conducts research that bridges his education and experience in analytical chemistry, aerosol science, and toxicology. Dr. McDonald has experience in the aerosolization and vaporization of gases and particles for a wide range of applications. He has an interest in developing laboratory exposures that represent "real world" conditions, and conducting characterizations of these exposures that allow toxicity results to be placed in the context of human exposures to entire environmental pollutants or drug products. His work spans the study of complex mixtures, respiratory drug delivery, animal model development, and metabolism in mammals. He is a member of the American Association for Aerosol Research, the Society of Toxicology, and the American Chemical Society. Dr. McDonald served on the NRC Committee to Review the Army's Enhanced Particulate Matter Surveillance Project Report. He earned a Ph.D. in environmental chemistry and toxicology from the University of Nevada.

Gunter Oberdörster, D.V.M., Ph.D., is a professor in the Department of Environmental Medicine at the University of Rochester, director of the University of Rochester Ultrafine Particle Center, principal investigator on a multidisciplinary research initiative in nanotoxicology, and head of the Pulmonary Core of a National Institute of Environmental Health Sciences center grant. His research focuses on the effects and underlying mechanisms of lung injury induced by inhaled nonfibrous and fibrous particles, including extrapolation modeling and risk assessment. His studies of ultrafine particles influenced the field of inhalation toxicology, raising awareness of the unique biokinetics and toxic potential of nanoscale particles. He has served on many national and international committees and is a recipient of several scientific awards. He is on the editorial boards of the *Journal of Aerosol Medicine*, *Particle and Fibre Toxicology*, *Nanotoxicology*, and the *International Journal of Hygiene and Environmental Health* and is the associate editor of *Inhalation Toxicology* and *Environmental Health Perspectives*. Dr. Oberdörster has served on several NRC committees, including the Committee for Review of the Federal Strategy to Address Environmental, Health, and Safety Research Needs for Engineered Nanoscale Materials and the Committee on Research Priorities for Airborne Particulate Matter. He earned his D.V.M. and Ph.D. (in pharmacology) from the University of Giessen, Germany.

Dorothy E. Patton, Ph.D., J.D., has more than 24 years experience with the U.S. Environmental Protection Agency (EPA) (1976–2000). She began her EPA career working as an attorney in EPA's Office of General Council on air, pesticide, and toxic substance issues. She later moved on to positions as Director of the Office of Science Policy, Executive Director of the EPA Science Policy Council, and Executive Director of the EPA Risk Assessment Forum. In these roles she was responsible for developing and implementing risk assessment policies and practices, environmental research planning and prioritization, and long-range strategic planning. After retiring from EPA in 2000, Dr. Patton taught a course in risk assessment at the Georgetown University Public Policy Institute and worked as a consultant with the Risk Sciences Institute within the International Life Sciences Institute. Dr. Patton was formerly a member of the NRC Board on Environmental Studies and Toxicology. She has also been a member of several NRC committees, including Human Biomonitoring for Environmental Toxicants, Metrics for Global Change Research, EPA Assessment Factors for Data Quality, Review of NASA's Earth Science Applications Program Strategic Plan, and the NRC Committee on Improving Risk Analysis Approaches Used by the United States EPA. She is currently a member of the IOM Committee on Decision-Making Under Uncertainty. Dr. Patton received a Ph.D. in Developmental Biology from the University of Chicago and a J.D. from the Columbia University School of Law.

William M. Valentine, Ph.D., D.V.M., is associate professor of pathology and researcher at the Vanderbilt University Medical Center. His research specializes in the mechanisms of environmental neurotoxicants and neurodegenerative disease. Dr. Valentine's current research includes delineating the molecular mechanisms of chemical agents that produce peripheral neuropathies. He is also investigating the role of copper dysregulation in neurodegenerative disease. Dr. Valentine has served on several NRC and IOM committees, including the Committee on Gulf War and Health: Literature Review of Selected Environmental Agents, Pollutants, and Synthetic Chemical Compounds, the Committee on Gulf War and Health: Review of the Literature on Pesticides and Solvents: Solvent Panel, and the Subcommittee on Jet Propulsion Fuel 8. Dr. Valentine earned his Ph.D. from the University of Illinois, Chicago, Illinois, and his D.V.M. from the University of Illinois, Champaign, Illinois.

Bailus Walker, Ph.D., M.P.H., is a professor of environmental and occupational medicine and toxicology at Howard University College of Medicine. His research interests include lead toxicity and environmental carcinogenesis. Dr. Walker has served as commissioner of public health for the Commonwealth of Massachusetts; chairman of the Massachusetts Public Health Council; and state director of public health for Michigan. He is past president of the American Public Health Association, and a distinguished fellow of both the Royal Society of Health and the American College of Epidemiology. Dr. Walker is currently a senior science advisor for environmental health to the National Library of Medicine and he is a member of the IOM. He has also served on several NRC committees, most recently the Committee on Improving Risk Analysis Approaches Used by the U.S. EPA; the Committee on Mine Placement of Coal Combustion Wastes; the Committee on Toxicology; and the Committee for the Review of the Army's Enhanced Particulate Matter Surveillance Program Report. Dr. Walker received a Ph.D. in occupational and environmental medicine from the University of Minnesota.

Appendix B

Review of Air Monitoring Data from Joint Base Balad

This appendix provides supplemental information on the air monitoring data collected at Joint Base Balad (JBB) and discussed in Chapter 4. The main focus of the appendix is on the strengths and limitations in the sample collection and analysis. A discussion of the sampling issues for specific chemical classes detected at JBB—polycyclic aromatic hydrocarbons (PAHs), particulate matter (PM), metals, volatile organic compounds (VOCs), and polychlorinated dibenzo-para-dioxins/furans (PCDDs/Fs)—is also given.

STRENGTHS AND LIMITATIONS OF THE AMBIENT DATA

In each monitoring campaign at JBB, ambient air concentrations of PM_{10}, metals, speciated VOCs, PAHs, and PCDDs/Fs were determined using standard EPA methodologies. In the 2009 campaign, $PM_{2.5}$ was also measured. The following standard EPA sampling methods for toxic organics (TO) were used (EPA 1990):

- TO-9 sampling for PCDDs/Fs;
- TO-13A sampling for PAHs;
- TO-14 sampling for VOCs in 2007;
- TO-15 sampling for VOCs in 2009; and
- MiniVol sampling for particulate matter with aerodynamic diameter less than 10 mm (PM_{10}) in 2007 and also less than 2.5 mm ($PM_{2.5}$) in 2009; the PM_{10} samples were also analyzed for individual metals by inductively coupled plasma mass spectrometry (ICPMS; Method 200.8).

For each method, samples are obtained by passing ambient air at a constant rate through a sampling apparatus for a fixed time, either to collect the materials (generally air pollutants) on a filter and/or sorbent (TO-9, TO-13A, Mini-Vol methods) or to collect a sample of the air itself (TO-14, TO-15 methods). The TO-9, TO-13A, TO-14, and TO-15 methods are designed to capture both vapor-phase and particulate-phase material. Photographs of the monitoring equipment are given in the CHPPM report and each instrument appears to have collected samples at about 4 feet above the ground (Taylor et al. 2008). The collected material is subsequently analyzed for specific chemicals. The measurements at JBB are the average concentration of the measured material in the air over the (approximately) 24-hour sampling period. Any spikes in concentration that may occur because of variations in source conditions or meteorology during the measurement period are incorporated into the 24-hour average, and are not discernible in the data except insofar as they affect the average.

The measurements also include all amounts of the material in ambient air regardless of the source. They do not directly measure emissions from the burn pit or from any other particular source. There are likely to be multiple sources of all the measured chemicals either on or in the vicinity of JBB, and the measurements include background concentrations due to those other sources as well as concentrations from the burn pit. The background concentrations from those other sources vary between sample times as a result of variations in time, number of sources, or location of those sources, as well as variations in meteorological conditions.

Several details of the sampling design and methodology affect the committee's ability to analyze the sampling data. Such details have different effects on different analyses, so the following comments should not be interpreted as criticisms of the sampling design or methodology, since the samples were collected for a different purpose.

- Samples were not collected simultaneously at all locations for each sampling event. An effort was made to sample all the individual pollutant types (VOCs, semivolatiles, PM, PCDDs/Fs) on the same day, and there is substantial overlap in the sampling start times and the length of time for each sample, but they vary by as much as several hours both within and between pollutant types. This limits the comparability of samples, since the weather conditions and sources of pollutants are likely to be different in character, number, and strength at different times of day. Consequently, comparisons between the different sites are made more difficult, and identification of distinct sources from the sample data is compromised to some extent.
- Samples were not collected on any planned schedule. Instead a "convenience sampling" approach was used in which samples were collected when logistically possible. As a result, averages of the sample results may be unrepresentative of long-term average exposures because of unrecognized patterns in exposure (e.g., daily, weekly, monthly, or seasonal variations) not compensated by averaging the measurements.
- The PM methodology used is not suited for collecting samples with high concentrations of PM, such as occur during dust storms, or even on many non-dust-storm days at JBB (NRC 2010). As a result, some of the filter samples probably suffer from sampling artifacts (underestimates or overestimates of PM concentration). The 2009 samples, where PM_{10} and $PM_{2.5}$ samples were collected simultaneously at the same locations, show evidence of such problems. The ratio of $PM_{2.5}$ to PM_{10} was between 0.25 to 0.6 for most samples, but a few showed substantially higher or lower ratios. In particular, for some samples the ratio exceeded unity (the ratio should always be less than unity except for measurement errors, since $PM_{2.5}$ is a component of PM_{10}). Similar problems are likely to have occurred with PM_{10} sampling in 2007, and may have been masked in some 2009 samples by occurring simultaneously in both PM_{10} and $PM_{2.5}$ samples. Further evidence of these problems is provided by some large discrepancies in measurements of the same parameter ($PM_{2.5}$ or PM_{10}) at the different sampling sites on the same day.
- Monitoring at different areas of the site would have been helpful. A site closer to the burn pit might have been more strongly and clearly affected by this source, allowing better characterization of the burn pit emissions. Sites in all the housing areas would allow better evaluation of exposures to all personnel on base rather than just those initially thought to be most highly exposed. Use of paired sites relatively close together, rather than single sites, may have compromised the analytic methods the committee attempted to use, since the paired sites were sufficiently far apart to be affected differently by local sources. Moreover, since these closely spaced sites were not sampled simultaneously, they do not provide any useful information on the variation of exposures within the housing area.
- The TO-14/TO-15 methodology (in which a sample of air is collected in an initially evacuated stainless steel flask) is not well suited to measure polar or reactive organics such as acrolein and 1,3-butadiene. Such materials may react on the walls of the stainless steel flask or in the gas phase during storage and transport of the sample. For example, 1,3-butadiene will decompose inside the canister during storage, mostly by reactions with nitrogen oxides. For risk assessments this is an important issue, because acrolein and butadiene are often the major risk contributors in screening risk assessments.

TABLE B-1 Average PAH Concentrations (ng/m^3) at JBB

	Guard Tower/ Transportation Field			H-6 Housing/CASF			Mortar Pit		
Analyte	Spring 2007	Fall 2007	2009	Spring 2007	Fall 2007	2009	Spring 2007	Fall 2007	2009
Acenaphthene	3.3	5.6	1.8	5.6	3.0	2.2	3.2	2.4	2.0
Acenaphthylene	5.7	24.4	3.5	12.1	14.2	8.3	4.9	8.7	2.7
Anthracene	2.8	6.2	1.3	3.4	2.8	1.0	2.0	2.0	0.5
Benz[a]anthracene	1.6	2.8	0.9	1.7	1.4	0.7	1.6	0.8	0.6
Benzo[a]pyrene	1.0	2.7	1.0	0.9	1.7	1.0	0.9	1.7	1.0
Benzo[b]fluoranthene	2.8	5.5	1.9	2.2	3.0	1.9	2.8	2.2	1.8
Benzo[e]pyrene	1.6	3.0	1.1	1.2	1.5	1.1	1.5	1.2	1.0
Benzo[g,h,i]perylene	1.5	3.3	1.3	1.2	2.7	1.7	1.4	1.7	1.4
Benzo[k]fluoranthene	0.6	1.2	0.5	0.5	0.7	0.4	0.6	0.5	0.5
Chrysene	2.5	3.7	2.5	2.4	2.3	1.6	2.1	1.3	1.4
Dibenz[a,h]anthracene	0.3	0.7	0.3	0.2	0.3	0.1	0.3	0.3	0.1
Fluoranthene	7.8	10.9	4.8	7.9	6.6	4.5	5.5	4.3	4.1
Fluorene	12.3	24.1	9.3	16.1	11.7	8.0	8.7	8.1	6.6
Indeno[1,2,3-cd]pyrene	1.4	3.2	1.1	1.1	2.0	1.2	1.3	1.7	1.3
Naphthalene	200.0	536.9	205.3	242.9	348.4	283.8	133.7	335.5	201.9
Phenanthrene	29.4	45.1	19.5	34.6	23.2	17.4	20.5	15.8	15.1
Pyrene	6.4	9.1	3.4	6.9	6.1	3.5	4.6	3.6	2.9
Number of samples	9	10	19	11	7	17	10	6	18

- The sampling locations were chosen to evaluate concentrations downwind of the burn pit (Taylor et al. 2008). The committee was initially informed that no military personnel serviced the burn pit at JBB, but subsequent information indicated that military personnel worked at or very near the pit. The concentration of contaminants in or near the pit cannot be inferred from any of the samples obtained, so there could be a subpopulation of military personnel who were highly exposed but whose exposure cannot be estimated from available data.

Polycyclic Aromatic Hydrocarbons

Measurements of PAHs were obtained at five sampling sites at JBB, but for 2007 these were reported as three locations—the guard tower/transportation field (20 samples),[1] H-6 housing/CASF (18 samples) and mortar pit (16 samples) (Table B-1). In 2007, samples were collected on each of 22 days, but on only 12 days were all three locations sampled. The 2009 measurements were again taken at the five sampling sites, and were reported separately for the five sites; but they are treated here as the three sampling locations as was done for the 2007 samples—the guard tower/transportation field (19 samples), H-6 housing/CASF (17 samples) and mortar pit (18 samples). Some samples were collected on each of 20 days, but only on 15 days were all three locations sampled. The nominally 24-h samples were not obtained simultaneously at the three locations on the common sampling days, but began within 2.5 hours in 2007, and within 1.6 hours in 2009; the sample times varied from 17.4 hours to 25.4 hours in 2007, and 21.1 hours to 25.2 hours in 2009. In both 2007 and 2009 each sample was analyzed for 17 PAH analytes, with all of the analytes detected in most samples.[2]

---

[1]Twenty-one samples results are reported in the data provided to the committee and used by Taylor et al. (2008), but one clearly corresponds to an unexposed sample and is omitted from further consideration here.

[2]The omitted sample was nondetect for every PAH except naphthalene, which was measured at a level 50 times lower than the next lowest sample. The committee considers this to be an unexposed sample.

Particulate Matter and Metals

The 2007 PM measurements included 28 to 32 samples at each of the three sampling locations (total 90 samples) on a total of 32 sampling days. PM_{10} samples were obtained by specifically collecting particles with aerodynamic diameters less than 10 mm on filters. The measurement consists of a careful weighing of the filter before and after the collection period; then knowledge of the amount of air directed through the filter allows computation of the average concentration of PM_{10} in the air. The PM on the filter was then analyzed for the following metals: antimony, arsenic, beryllium, cadmium, chromium, lead, manganese, nickel, vanadium, and zinc. For the CHPPM samples at JBB in 2007, the measurements of individual metals were not useful, because the detection limit of the method used was too high to detect the metals of interest in the PM_{10} material in the great majority of samples. Only 6 detections, all of lead, were made in the 90 samples; these detections were all made on 2 consecutive days in November 2007 and all were less than 0.7 mg/m^3. The committee therefore disregarded the metals measurements as not providing useful information.

In 2009, 50 measurements were available for 19 days of near-simultaneous PM_{10} or $PM_{2.5}$ samples, together with another 8 samples of PM_{10} or $PM_{2.5}$ individually, although four of the PM_{10} measurements were clearly affected by measurement artifacts and were ignored by the committee. The collected PM_{10} material was analyzed for metals, and again the detection limits were sufficiently high that most of the samples had nondetectable levels of the metals of interest. There were, however, many more detections than in 2007—25 of 108 samples had at least one metal detected—despite detection limits that were very similar; the concentrations of the metals varied, with individual concentrations up to 5 mg/m^3. The third highest PM_{10} measurement was discounted because it suffered from measurement artifact (by comparison with the simultaneous $PM_{2.5}$ measurement), and the consistency of the two available $PM_{2.5}$ measurements on the same day strongly suggests that the second highest PM_{10} measurement also suffered from the same problem.[3]

Volatile Organic Compounds

VOCs were detected in 66 samples in 2007 and in 55 samples in 2009. The likely sources were considered to be combustion (including petroleum fuel combustion), fuel additives, solvents, and refrigerants. The 2007 measurements of VOCs included 66 samples on 26 days, using TO-14 methodology for sample collection. Each sample was analyzed for 78 VOCs. Fifty-five of these analytes were detected in six or fewer samples. The frequency of detection and likely major sources or uses for the other 28 analytes are shown in Table B-2.

The 2009 measurements included 57 samples on 20 days. There was a change to TO-15 methodology, so each sample was analyzed for up to 62 VOCs, including some not measured in 2007, and some of those measured in 2007 were omitted from the analyte list in 2009. Forty of the VOCs were detected in six or fewer samples. The frequency of detection and likely major sources or use for the other 22 analytes are shown also in Table B-2.

The differences between these lists largely arise from the different analytes measured or different detection limits in 2007 and 2009. Octane, isooctane, chlorodifluoromethane and pentane were analyzed in 2007 but not in 2009, while the xylenes were analyzed in different combinations. Isopropyl alcohol and cyclohexane were analyzed in 2009 but not in 2007. The detection limits for acrolein, 2-butanone, methylene chloride, and 1,4-dichlorobenzene were lower in 2009 than in 2007, although as discussed earlier the method used for acrolein was inadequate. The detection limit for MtBE was higher in 2009 than in 2007, and the average detected concentration in 2007 was half the 2009 detection limit. However, 4-ethyltoluene was detected less frequently and at lower concentrations in 2009 (even with a lower detection limit).

Table B-3 presents average concentrations of the twelve most frequently detected VOCs by location and sampling campaign, with nondetects assumed to contribute one-half the detection limit (for these VOCs, setting nondetects to zero alters the mean estimate by a factor of 1 to 2.4). Like PM_{10}, VOC concentrations were similar

---

[3]This assumes negative artifacts (underestimation of concentrations). Since positive artifacts are also quite likely at high PM concentrations with the methodology used for measurement at JBB (NRC 2010), this evaluation of artifact could be entirely backward. In that case the majority of high concentration measurements could be artifacts.

TABLE B-2 Number of Detects and Likely Source for Analytes Detected More than Occasionally in 2007 and 2009

Analyte	Number of detects[a] 2007	Number of detects[a] 2009	Likely major source or use
Acetone	66	55	Solvent, combustion
Benzene	66	56	Combustion
Chloromethane	66	54	Natural sources (ATSDR 1998, 1999)
Toluene	63	56	Petroleum-based fuel combustion
Hexane	65	53	Petroleum-based fuel combustion
Pentane	65	NA	Petroleum-based fuel combustion
Dichlorodifluoromethane	64	55	Refrigerant
Propylene	32	55	Petroleum-based fuel combustion
2-Butanone (MEK)	20	53	Solvent, combustion
Methylene chloride	27	52	Solvent
o-Xylene	51	52	Combustion
n-Heptane	60	36	Petroleum-based fuel combustion
Ethylbenzene	49	49	Combustion
m,p-Xylene	55	NA	Combustion
Xylenes, total	NA	45	Combustion
1,2,4-Trimethylbenzene	51	41	Combustion
Octane	46	NA	Petroleum-based fuel combustion
Trichlorofluoromethane	42	37	Refrigerant
Chlorodifluoromethane	41	NA	Refrigerant
Acrolein	4	34	Combustion
4-Ethyltoluene	38	12/30	Combustion
Isopropyl alcohol	NA	27	Solvent, disinfectant
Methyl tert-butyl ether (MtBE)	29	0	Anti-knock fuel additive
1,3,5-Trimethylbenzene	24	11	Combustion
Isooctane	17	NA	Anti-knock fuel additive
Styrene	17	14	Combustion
1,4-Dichlorobenzene	4	9	Mothballs, pesticide
Cyclohexane	NA	8	Petroleum-based fuel combustion

NOTE: NA = not analyzed.
[a]Total possible detects 66 in 2007 and 57 in 2009 except 4-ethyltoluene.

for many analytes at all the measurement locations at JBB, and there did not appear to be any consistent gradients in concentration, although differing gradients exist for some analytes at some times.

Polychlorinated Dibenzo-Para-Dioxins/Furans

During 2007, 18, 21, and 21 PCDD/F nominal 24-hour samples were collected at the guard tower/transportation field, H-6 housing/CASF, and mortar pit locations, respectively, on 20 sampling days (TO-9 method; 60 samples total). Two of the samples for the H-6 housing/CASF and mortar pit locations were obtained on the same day (about 2 hours and 20 minutes apart in starting time, respectively), and on 2 of the 20 sampling days the guard tower/transportation field was not sampled. All samples were analyzed for the seventeen 2,3,7,8-chlorinated PCDD/F congeners, with all congeners detectable in 41 or more of the 60 samples except 1,2,3,7,8,9-hexaCDF (detected in 14/60 samples).

The 2009 sampling data for PCDD/PCDFs included 19, 17, and 18 PCDD/F nominal 24-hour samples at the

TABLE B-3 Average Concentration (mg/m^3) of the 12 Most Frequently Detected VOCs at JBB

Analyte	Guard Tower/ Transportation Field			H-6 Housing/CASF			Mortar Pit		
	Spring 2007	Fall 2007	2009	Spring 2007	Fall 2007	2009	Spring 2007	Fall 2007	2009
Benzene	6.8	9.1	4.9	4.8	5.0	3.8	3.3	8.3	2.7
Acetone	37.9	10.8	29.6	25.7	9.4	42.7	13.8	6.8	29.7
Chloromethane	2.0	1.6	1.9	1.9	1.1	1.7	1.5	1.2	1.8
Dichlorodifluoromethane	4.3	2.5	2.5	3.0	2.6	2.6	2.7	2.3	2.5
Toluene	11.8	23.0	10.7	8.6	30.2	13.1	5.0	55.2	9.8
Hexane	5.1	4.5	5.3	2.9	5.4	31.6	2.1	16.2	7.2
Xylenes	7.2	22.4	7.7	9.6	18.3	8.9	4.6	52.6	6.4
Ethylbenzene	4.0	4.8	2.9	3.2	5.3	2.7	2.4	9.6	1.7
n-Heptane	3.6	2.9	2.1	4.1	3.5	2.8	1.8	6.5	1.9
1,2,4-Trimethylbenzene	2.6	18.0	2.0	3.7	9.3	2.9	2.2	11.9	1.8
Propylene	2.7	3.8	2.2	1.6	2.8	2.0	1.2	4.8	1.3
Methylene chloride	10.1	2.7	17.3	2.3	2.1	17.9	2.5	3.0	8.3
Number of samples	15	8	19	12	9	18	15	7	19

guard tower/transportation field, H-6 housing/CASF, and mortar pit locations, respectively, on 19 sampling days, with 3 different days missing a sample at one location (TO-9 method; 54 samples total). The concentrations measured in May and June of 2009 were consistently lower for all congeners, and at all measurement locations, than those measured in January, February, April, October, and November 2007, with a larger proportion of nondetects than in 2007 (206/918, 22%, versus 126/1173, 11%). Again, 1,2,3,7,8,9-hexaCDF was the most frequent nondetect (only 4/54 samples measurable).

CHPPM also provided the committee with PCDD/PCDF sample data for 9 samples collected on 9 dates in 2006, but no location for these samples was given. At JBB, the distributions of concentrations of total PCDD/Fs (the sum of the seventeen 2,3,7,8-chlorinated congener concentrations) and individual congeners at each location are approximately lognormal with high correlations between congeners. In the 2007 measurements, there is no significant difference in individual congeners or in total PCDD/F between the January–April and October–November measurements at each location, although there is a slight trend for the January–April concentrations to be higher than the October–November concentrations at the guard tower/transportation field and H-6 housing/CASF locations.

REFERENCES

ATSDR (Agency for Toxic Substances and Disease Registry). 1998. Toxicological profile for chloromethane. *Tox profiles*. Atlanta, GA: Agency for Toxic Substances and Disease Registry. http://www.atsdr.cdc.gov/toxprofiles/tp.asp?id=587&tid=109 (accesed March 3, 2011)

EPA (U.S. Environmental Protection Agency). 1990. *Technical assistance document for sampling and analysis of toxic organic compounds in ambient air*. Washington, DC: Atmospheric Research and Exposure Assessment Laboratory.

NRC (National Research Council). 2010. *Review of the Department of Defense Enhanced Particulate Matter Surveillance Program report*. Washington, DC: The National Academies Press.

Taylor, G., V. Rush, A. Peck, and J. A. Vietas. 2008. *Screening health risk assessment burn pit exposures Balad Air Base, Iraq and addendum report*. IOH-RS-BR-TR-2008-0001/USACHPPM 47-MA-08PV-08. Brooks City-Base, TX: Air Force Institute for Operational Health and U.S. Army Center for Health Promotion and Preventative Medicine.

Appendix C

Epidemiologic Studies Cited in Chapter 6: Health Outcomes

Table C-1 includes descriptions of epidemiologic studies cited in Chapter 6. Key and supporting studies are presented alphabetically with studies of Gulf War veterans listed in a separate section at the end of the table. Text in italics reflects the section and designation (k = key, s = supporting) of each study as it is cited in the chapter.

TABLE C-1 Epidemiologic Studies Cited in Chapter 6: Health Outcomes

Study and Design	Population	Exposure	Outcomes	Adjustments	Limitations and Comments
Aronson et al. 1994 Retrospective cohort study. *[Respiratory-s; Circulatory-k; Cancer-k; All Cause-k]*	5,414 firefighters who had worked at any time between 1950 and 1989 at six fire departments in Metropolitan Toronto. Includes 777 deaths.	Employed as a firefighter (yes/no); years since first exposure (first employment); years of exposure (years of employment).	Cause of death. Compared with general male population of Ontario. Total Cohort: Brain cancer and other NS tumors (ICD9 191-192): SMR 201, 95% CI 110-337; Other Malignant Neoplasms (ICD9 195-199): SMR 238, 95% CI 145-367; Diabetes mellitus (ICD9 250): SMR 35, 95% CI 9-88; Chronic rheumatic heart disease (ICD 393-398, 424.0-424.3): SMR 15, 95% CI 0.4-85; Aortic aneurysm (ICD9 441): SMR 226, 95% CI 136-354; Symptoms/Ill-defined (ICD9 780-799): SMR 17, 95% CI 0.4-95; External Causes (ICD9 E800-999): SMR 71, 95% CI 55-90). Age 60-84: Aortic aneurysm: SMR 245, 95% CI 140-398; Chronic bronchitis, asthma, and emphysema: 155, 95% CI 101-227; Digestive system diseases: SMR 156, 95% CI 100-232; Gallbladder diseases: 420, 95% CI 136-980; all other causes NS.	Modified life-table approach; subjects censored at age 85 years; cata stratified by years since first employment, age, and duration of employment.	Multiple testing-findings labeled as statistically significant by chance alone; healthy worker effect and survivor effect.
Aronson et al. 1996 Registry-based case-control study *[Reproductive-s]*	9,340 Fathers of all children with congenital heart defects matched to 9,340 Toronto fathers of children without an anomaly, from the Toronto birth registry 1979-1986.	Father's occupation as a firefighter in Toronto.	Cardiac congenital anomalies. 11 case and 9 controls had fathers who were firefighters (OR 1.22, 95% CI 0.46-3.33).	Matched on birth year, maternal age at birth, birth order, parents' birth places (in or out of Ontario), and mother's marital status at birth.	
Bandaranayke et al. 1993 Case-control study *[Neurological-s]*	245 firefighters exposed to a chemical fire in 1984, and 217 unexposed firefighters matched for age and years of service in New Zealand.	Presence at a chemical fire in 1984.	Nervous system dysfunction. More exposed firefighters exhibited CNS dysfunction than unexposed firefighters (in 4 categories of symptoms, all p<0.025). Firefighters score poorly on more neuropsychological tests (RR 1.32, 95% CI 1.11-1.57), particularly psychomotor tests (RR 1.51, 95% CI 1.33-1.71, compared to unexposed firefighters.	Matched for age and years of service.	Tests conducted 4 years after the fire event; 3 cases of testicular cancer described.

Baris et al. 2001 Retrospective cohort study [Respiratory-s; Neurological-k; Circulatory-k; Cancer-k; All Cause]	7,789 Firefighters in Philadelphia employed between 1 Jan. 1925, and 31 Dec. 1986. Females excluded. 2,220 deaths.	Duration of employment, company runs, and station house design were used as a surrogate for individual exposure. Categories of runs were categorized as low, medium, high.	No differences in hospital admissions, health problems, prevalence of allergies, history of miscarriages due to abnormality or stillbirth, or birth defects, psychological histories, tobacco or alcohol use, abnormal ECGs, blood cell counts were detected between the exposed and unexposed firefighters. Cause of death compared with the general US white male population; RR between high and low exposure groups. Total Cohort: All causes: SMR 0.96, 95% CI 0.92-0.99; All cancers: SMR 1.10, 95% CI 1.00-1.20; Colon cancer: SMR 1.51, 95% CI 1.18-1.93; Ischemic heart disease: SMR 1.09, 95% CI 1.02-1.16; Cerebrovascular Disease: SMR 0.83, 95% CI 0.69-0.99; Respiratory diseases: SMR 0.67, 95% CI 0.55-0.82; Genitourinary diseases: SMR 0.54, 95% CI 0.36-0.81; External causes of death: SMR 0.69, 95% CI 0.59-0.80; All accidents: SMR 0.72, 95% CI 0.59-0.86; Suicide: SMR 0.66, 95% CI 0.48-0.92; All other causes NS. ≤ 9 Years Employment: All cancers: SMR 1.26, 95% CI 1.07-1.49; Colon cancer: SMR 1.78, 95% CI 1.12-2.82; Lung cancer: SMR 1.52, 95% CI 1.16-2.01; Pancreatic cancer: SMR 2.33, 95% CI 1.36-4.02; Prostate cancer: SMR 2.36, 95% CI 1.42-3.91; Genitourinary disease: SMR 0.27, 95% CI 0.10-0.71; External causes of death: SMR 0.61, 95% CI 0.49-0.77; All accidents: SMR 0.56, 95% CI 0.43-0.75; All other causes NS. 10-19 Years Employment: Circulatory diseases: SMR 1.20, 95% CI 1.10-1.31; Ischemic heart disease: SMR 1.35, 95% CI 1.21-1.49; Respiratory diseases: SMR 0.68, 95%CI 0.49-0.96; Suicide: SMR 0.37, 95% CI 0.18-0.78; All other causes NS.	Age and calendar-year adjusted with a 10-year lag period; stratified by position, duration of employment, age at risk, hire period, company type (ladder, engine or both).	Duration of employment, company runs and station house design as a surrogate for individual exposure misclassification. Healthy worker effect and survivor effect.

continued

TABLE C-1 Continued

Study and Design	Population	Exposure	Outcomes	Adjustments	Limitations and Comments
			≥ 20 Years Employment: All causes: SMR 0.91, 95% CI 0.85-0.98; Colon cancer: SMR 1.68, 95% CI 1.17-2.40; Kidney cancer: SMR 2.20, 95% CI 1.18-2.49; Multiple myeloma: SMR 2.31, 95% CI 1.04-5.16; Benign neoplasms: SMR 2.54 1.06-6.11; Circulatory diseases: SMR 0.90, 95% CI 0.82-0.99; Respiratory diseases: SMR 0.59, 95% CI 0.42-0.82; Emphysema: SMR 0.39, 95% CI 0.16-0.93; All other causes NS.		
Bates 1987 Cohort study [Circulatory-s]	596 men who worked for 6 yrs or more in the Toronto Fire Department; hired from 1949-1959.		Comparing High to Low (> or ≤ 3,191 cumulative runs) Exposure: All causes: RR 0.81, 95% CI 0.72-0.92; Buccal cavity and pharynx cancer: RR 0.19, 95% CI 0.04-0.96; Circulatory diseases: RR 0.78, 95% CI 0.65-0.93; Ischemic heart diseases: RR 0.77, 95% CI 0.63-0.95; External causes of death: RR 0.61, 95%CI 0.39-0.95; All other causes NS. Cardiovascular mortality ages 45 to 54 though 1984, compared to mortality rates of Toronto. 52 deaths from all causes, 21 from coronary artery diseases. Ages 45-49: SMR 1.80, 95% CI 1.01-3.19 Ages 50-54: SMR 1.75, 95% CI 0.90-3.39 Ages 45-54: SMR 1.73, 95% CI 1.12-2.66.	Standardized by age, sex, and calendar year.	
Bates 2007 Registry-based case-control study [Cancer-k]	3,659 firefighters with cancer and 800,448 non-firefighter controls with cancer, in from the California Cancer Registry, 1988-2003.	Ever employed as firefighter.	Cancer diagnosis among firefighters compared to cancer diagnoses among other occupations. Esophageal: OR 1.48, 95% CI 1.14-1.91; Melanoma skin: OR 1.50, 95% CI 1.33-1.70; Prostate: OR 1.22, 95% CI 1.12-1.33; Testicular: OR 1.54, 95% CI 1.18-2.02; Brain: OR 1.35, 95% CI 1.06-1.72; All other sites NS.	Age, calendar period of diagnosis, race, socio-economic status (by census block of residence).	
Bates et al. 2001 Retrospective cohort study	4,305 firefighters employed in New Zealand between 1977-1995.	Ever employed as a firefighter; duration of employment as a firefighter.	Cancer incidence and mortality, calculated as SIR and SMR relative to New Zealand male population, follow-up 1977 through 1995 for mortality, 1996 for cancer. Cancer incidence 1977-1996: All sites NS.	Age, sex and calendar period standardized.	This study follows up on a testicular cancer cluster described by Bandaranayake et al. 1993.

[Cancer-s]			Cancer incidence 1990-1996: Testicular cancer incidence: SIR 2.97, 95% CI 1.3-5.9; all other sites NS. Mortality 1977-1995: All causes: SMR 0.58, 95% CI 0.5-0.7; Circulatory diseases: SMR 0.54, 95% CI 0.4-0.7; Ischemic heart disease: SMR 0.58, 95% CI 0.4-0.8; External causes: SMR 0.69, 95% CI 0.3-0.8; All other causes NS.		Possibly confounded by high level of awareness of testicular cancer in this population. Healthy worker effect.
Beaumont et al. 1991 Retrospective cohort study. [Respiratory-s; Neurological-s; Circulatory-k; Cancer-k; All Cause]	3,066 white male firefighters from San Francisco employed 1940-1970. 1,186 deaths.	Firefighter employment (yes/no). Length of employment.	Cause of death. Compared with the general US male population. All causes: RR 0.90, 95% CI 0.85-0.95; Tuberculosis: RR 0.26, 95% CI 0.07-0.68; Diabetes mellitus: RR 0.36, 95% CI 0.14-0.75; Diseases of the heart: RR 0.89, 95% CI 0.81-0.97; Respiratory diseases: RR 0.63, 95% CI 0.47-0.83; Acute respiratory infections: RR 0.63, 95% CI 0.40-0.95; Emphysema: RR 0.52: 0.24-0.99; Digestive system diseases: RR 1.57, 95% CI 1.27-1.92; Cirrhosis and other liver diseases: RR 2.27, 95% CI 1.73-2.93; Accidental falls: RR 1.9, 95% CI 1.18-2.91; Cancer of digestive organs and peritoneum: RR 1.27, 95% CI 1.04-1.55; Esophageal cancer: RR 2.04, 95% CI 1.05-3.57; Prostate cancer: RR 0.38, 95% CI 0.16-0.75. All other causes NS. 3-19 Years Since First Employment: All neoplasm sites NS. 20-29 Years Since First Employment: All cancer sites: RR 0.67, statistically significant (95% CI not reported); All other neoplasm sites NS 30-39 Years Since First Employment: Billiary passages, liver and gall bladder cancer: RR 3.87, statistically significant (95% CI not reported); All other neoplasm sites NS 40+ Years Since First Employment: Stomach cancer: RR 2.32, statistically significant (95% CI not reported); All other neoplasm sites NS.	Rate ratios standardized for age, year, sex, and race.	Reliability of death certificates. Potential healthy worker effect.

continued

TABLE C-1 Continued

Study and Design	Population	Exposure	Outcomes	Adjustments	Limitations and Comments
			3-9, 10-19, and 20-29 Years of Employment: All neoplasm sites NS. 30+ Years of Employment: Billiary passages, liver and gall bladder cancer: RR 3.87, statistically significant (95% CI not reported); All other neoplasm sites NS.		
Betchley et al. 1997 Cohort study *[Respiratory-s]*	Full-time and seasonal wildland fire management workers in Region 6 of USDA Forest Service and Bureau of Land Management in Salem during the 1992 season. 76 subjects were studied for cross-shift and 53 for cross-season analysis.	Shift or season of firefighting.	Spirometric measurements of lung function and self administered questionnaire data were collected before and after the 1992 firefighting season. Cross-season data were collected on average 77.7 days after the last occupational smoke exposure. Cross-season analysis: Mean individual declines for FVC ($p=0.28$), FEV_1 ($p=0.03$) and FEF_{25-75} ($p=0.02$) of 0.033 L, 0.104 L, and 0.275 L/sec, respectively; no significant difference in respiratory symptoms. Cross-shift analysis: The pre-shift to mid-shift decreases were 0.089 L, 0.190 L, and 0.439 L/sec, respectively; pre-shift to post-shift declines of 0.065 L, 0.150 L, and 0.496 L/sec (all $p<0.01$); no significant difference in respiratory symptoms.		
Biggeri et al. 1996 Case-control study *[Cancer-s]*	755 male lung cancer deaths and 755 controls from a local autopsy registry in Trieste, Italy.	Exposure model based on residential distance and direction from sources of pollution-shipyard, iron foundry, incinerator, or city center.	Excess relative risk of 6.7 at zero distance ($p<0.001$), with risk dropping rapidly with distance (slope = –0.176).	Adjusted for age, smoking, likelihood of occupational carcinogen exposures, levels of PM.	Lung cancer also significantly related to distance from the city center.
Bresnitz et al. 1992 Cross-sectional study *[Respiratory-s]*	86 male incinerator workers at a facility in Philadelphia employed in June 1988.	High or low exposure was determined by an industrial hygienist based on job description, duration, and data from personal breathing zone and	Spirometry, blood and urine samples, physical exams, questionnaires used to collect health, medical, and employment information. Elevated exposure was not significantly related to biomarkers of exposure, hypertension, proteinuria, or changes in pulmonary function.	Stratified by smoking and alcohol. No correction for multiple comparisons.	High and low exposure groups differed by duration of employment and alcohol intake; no unexposed comparison group.

145

Study	Design	Population	Exposure	Outcome	Adjustments	Comments
Burnett et al. 1994 Retrospective cohort study [Circulatory-s; Cancer-s]		5,744 white male firefighter deaths identified from National Occupational Mortality Surveillance system (includes 27 states) from 1984-1990.	general air sample for dioxins and furans. Ever employed as firefighter.	Cause of death Total Cohort All cancers: PMR 110, 95% CI 106-114; Rectal cancer: PMR 148, 95% CI 105-205); Skin cancer: PMR 163, 95% CI 115-223; Kidney cancer: PMR 144, 95% CI 108-189; Lymphatic and hematopoietic cancers: PMR 130, 95% CI 111-151; Non-Hodgkin's lymphoma: PMR 132, 95% CI 102-167; Multiple myeloma: PMR 148, 95% CI 102-207; Accidental falls: PMR 149, 95% CI 109-199; and Fire-related accidents: PMR 242, 95% CI 157-357. For deaths under age 65 All cancers: PMR 112, 95% CI 104-121; Rectal cancer: PMR 186, 95% CI 104-121; Skin cancer: PMR 167, 95% CI 107-248; Lymphatic and hematopoietic cancers: PMR 161, 95% CI 129-199; Non-Hodgkin's lymphoma: PMR 161, 95% CI 112-224; Leukemia: PMR 171, 95% CI 118-240; Accidental falls: PMR 206, 95% CI 129-312; and Fire-related accidents: PMR 335, 95% CI 157-357.	Age, race, sex adjusted.	
Calvert et al. 1999 Cohort study [Circulatory-s]		Deaths among 488,539 white males and 104,988 black males in the National Occupational Mortality Surveillance System (covers 27 states), 1982-1992.	Employment as a firefighter.	Ischemic heart disease deaths in males 16-60 for firefighting occupations. 434 white (PMR 104, 95% CI 94-114) and 26 black (PMR 169, 95% CI 110-247) deaths among firefighters due to ischemic heart disease.	Age standardized.	Differential reporting of ischemic heart disease may be affected by presumption of disease for certain occupational exposures (such as for firefighting); elevated PMRs reported for many other occupations.
Carozza et al. 2000 Population-based case-control study [Cancer-s]		476 cases of glioma in San Francisco adults diagnosed 1991-1994 and 462 controls.	Employment as a fireman.	Glioma incidence 3 cases and 1 control among those ever employed as a fireman (OR 2.7, 95% CI 0.3-26.1). No significant associations when stratified by latency, duration of employment, or tumor type.	Controlled for age, gender, education, and race.	Few cases and controls among firemen.

continued

TABLE C-1 Continued

Study and Design	Population	Exposure	Outcomes	Adjustments	Limitations and Comments
Charbotel et al. 2005 Longitudinal study [Respiratory-s]	83 incinerator workers and 76 age-matched non-exposed workers followed over 3 years. Workers categorized by exposure/task.	2 incinerators in urban areas of France. Air sampling performed, presented in an earlier publication (Maitre et al. 2003).	Spirometry measurements from 1st and 3rd years. Medical history and symptoms assessed by questionnaire. Baseline lung function lower in exposed workers than controls (not significant). In the third year, controls had significantly better percent predicted values for FEF_{50}/PV (p=0.04), FEF_{25-75}/PV (p=0.02), and FEF_{25-75}/FVC (p=0.01), but not for other measures of lung function. No decrease in lung function (first to third year) was seen related to exposure. After adjustment for smoking, medical history, and examination center, FEF_{25-75} in the 3rd year was lower in incinerator workers than in unexposed workers (mean±SD % predicted, 94.1±27.9 vs 105.5±25.3). No relationship between exposure and lung function change during follow-up.	Adjustment for smoking, history of allergy or lung disease, and examination center.	Follow-up of Hours et al. 2003.
Comba et al. 2003 Case-control [Cancer-s]	37 cases of soft tissue sarcoma in diagnosed 1989-1998 and residing in Mantua, Italy, and 3 neighboring communities compared to 171 randomly selected unexposed controls from the population matched for age and sex.	Residential distance from an industrial waste incinerator.	Soft tissue sarcoma incidence Less than 2 km: OR 31.4 (95% CI 5.6-176.1), based on 5 cases. Greater than 2 km: No significant increase from null. No significant decrease in risk observed with increasing distance from source when measuring continuously.	Matched for age and sex.	Few exposed cases.
Cordier et al. 2004 Ecological study [Reproductive-s]	Malformed children born to residents of 194 exposed communities surrounding incinerators compared to 2678 unexposed communities.	Exposure to incinerator emissions was estimated from a plume model. Interested in dioxin- and metal-contaminated PM.	Obstructive uropathies; cardiac, urinary, and skin anomalies identified through population-based birth defects registry and active search of medical records. Facial cleft (RR 1.30, 95% CI 1.06-1.59) and renal dysplasia (RR 1.55, 95% CI 1.10-2.20) more frequent in exposed populations, and other renal anomalies was lower (RR 0.44, 95% CI 0.20-0.97). A dose-response trend	Adjusted for year of birth, maternal age, department of birth, population density, average family income and, when available, road traffic.	Few measurements of total dusts, dioxins, and metals available; rates of cardiac anomalies, obstructive uropathies, and skin anomalies likely explained by road traffic density.

Study	Population	Exposure	Results	Comments
			was observed with increasing exposure for obstructive uropathies (p=0.07). Dose-response trends with increasing road traffic density were found for cardiac anomalies (p=0.02), skin anomalies (p=0.02), and obstructive uropathies (p=0.07).	Rate increases in later years hard to interpret without information on cumulative exposure or increases in exposure.
Cresswell et al. 2003 Ecological study [Reproductive-s]	1,508 cases from total 81,255 live births, stillbirths, induced abortions and fetal death after 14 weeks gestation to mothers residing within 7 km of a waste incinerator, 1985-1999 from the Northern Regional Congenital Abnormality Survey.	Based on distance from incinerator which went into operation in 1988, location within inner (3 km radius) or outer (3-7 km radius) areas and pre- vs. post-incinerator operation.	Chromosomal and non-chromosomal congenital anomalies comparing the inner and outer zones. No significant overall association between number of anomalies and residential proximity to incinerator was found. Risks were not elevated pre- or post-1988, but when stratified by year, the risk was significantly elevated in 1995 (OR 1.73, 95% CI 1.10-2.72), 1998 (OR 1.56, 95% CI 1.01-2.41), and 1999 (OR 2.05, 95% CI 1.20-3.52).	Adjusted for socio-economic deprivation and year.
Demers et al. 1992a Retrospective cohort study. [Respiratory-s; Circulatory-k; All Cause]	4,546 male firefighters in Seattle and Tacoma, WA, and Portland, OR, for at least a year, 1944-1979. 1,162 deaths	Employment as a firefighter (yes/no). Duration in fire combat positions.	SMR compared with the US white male population. IDR for cause of death compared with police officers from the same cities. All causes: SMR 0.81, 95% CI 0.77-0.86; Kidney cancer: SMR 0.27, 95% CI 0.03-0.97; Bladder and other urinary cancers: SMR 0.23, 95% CI 0.03-0.83; Brain and nervous system cancers: SMR 2.07, 95% CI 1.23-3.28; Heart diseases: SMR 0.79, 95% CI 0.72-0.87; Ischemis heart disease: SMR 0.82, 95% CI 0.74-0.90; All other causes NS. All causes: IDR 0.87, 95% CI 0.79-0.95; Other circulatory diseases: IDR 0.72, 95% CI 0.54-0.96; Cerebrovascular disease: IDR 0.65, 95% CI 0.45-0.92; All other causes NS. <10 Years of Employment: All causes NS. 10-19 Years of Employment: Brain and nervous system tumors: SMR 3.53, 95% CI 1.5-7.0; All other causes NS 20-29 Years of Employment: All causes NS.	Age and calendar-year standardized. Stratified by years of fire combat exposure, years since first employment, and age at risk. No adjustment for smoking or other potential confounders. Large study size. Disease misclassification dependent on accuracy of death certificates. Statistical instability when comparing firefighter mortality to police due to relatively few deaths among police.

continued

147

TABLE C-1 Continued

Study and Design	Population	Exposure	Outcomes	Adjustments	Limitations and Comments
			≥ 30 Years of Employment: Lymphatic and hematopoietic cancers: SMR 2.05, 95% CI 1.1-3.6; Leukemia: SMR 2.60, 95% CI 1.0-5.4; Diseases of arteries, veins, and pulmonary circulation: SMR 1.99, 95% CI 1.3-2.9; All other causes NS.		
			<20 and 20-29 Years Since First Employment: All causes NS. ≥30 Years Since First Employment: Prostate cancer: SMR 1.42, 95% CI 1.0-2.0; Brain and nervous system tumors: SMR 2.63, 95% CI 1.2-4.4; Lymphatic and hematopoietic cancers: SMR 1.48, 95% CI 1.0-2.2; Diseases of arteries, veins, and pulmonary circulation: SMR 1.33, 95% CI 1.0-1.8; All other causes NS.		
Demers et al. 1992b Retrospective cohort study [Cancer-k]	4,528 male firefighters and police officers in Seattle and Tacoma, WA, employed for at least 1 year between 1944 and 1979. 338 death certificate registry and 174 identified cancer cases.	Employment as a firefighter or police officer (yes/no).	SMR cancer deaths compared with the white male population of Washington State and SIR incident cancer cases compared with all males with malignancies in the same counties (follow-up 1945-1989). Cancer Incidence Prostate cancer: SIR 1.37, 95% CI 1.11-1.69; All others NS. Cancer Mortality Stomach cancer: SMR 2.04, 95% CI 1.05-3.56; All others NS.	Standardized by age and calendar year. No adjustment for smoking or other potential confounders.	Large study size. Analysis lumps firefighters and police together, limiting generalizeability.
Demers et al. 1994 Retrospective cohort study [Cancer-k]	2,447 male firefighters in Seattle and Tacoma, WA, employed for at least 1 between 1944 and 1979. 224 cancer cases among firefighters.	Employment as a firefighter (yes/no); Duration of active duty firefighting/ employment (years).	SIR and IDR Cancer incidence (1974-1989). Compared with mortality rates for the Seattle and Tacoma areas and local police officers. Firefighters compared with local cancer incidence rates. Prostate cancer: SIR 1.4, 95% CI 1.1-1.7; All others NS. Firefighters compared with police IDR for all cancer sites NS.	Adjusted for age and calendar-period; stratified by years since first employment and duration of employment; no adjustment for smoking or other potential confounders.	Cohort previously reported by Heyer et al. 1990 and Demers et al. 1992. Small numbers of police cancer cases limits the precision of risk estimates.
Deschamps et al. 1995 Prospective cohort study	830 male firefighters having served at least 5 years (as of Jan. 1, 1977) for the Brigade des sapeurs-	Time spent working on assignments involving active fire combat duty (as opposed to	Cause of death (1977-1991); compared with the general French male population. All cause mortality: SMR = 0.52 (0.35-0.75); All causes others NS.	Age and calendar adjusted SMRs; no adjustment for smoking or other potential confounders.	Healthy worker effect; few deaths.

Study	Population	Exposure	Results	Adjustments	Comments
[Respiratory-s; Circulatory-s; Cancer-s; All Cause]	pompiers de Paris (BSPP). Includes 32 deaths and 11,414 person-years.	office-work). This exposure was only determined for deceased persons and was evaluated from BSPP records.			
Dibbs et al. 1982 Longitudinal study *[Circulatory-k]*	171 male firefighters and 1,475 non-firefighters participating in the Normative Aging Study at the VA outpatient clinic in Boston, MA having completed three complete medical examinations.	Firefighter (yes/no).	Coronary heart disease, myocardial infarction, angina pectoris after 10 years of follow-up. Compared with non-firefighters. Coronary heart disease: IRR 0.5, 95% CI 0.2-1.4. Myocardial infarction: IRR 0.5, 95% CI 0.1-1.9.	Data on serum cholesterol, blood pressure, BMI, age, and cigarette smoking stratified by firefighter/non-firefighter to detect differences in risk factors for coronary heart disease.	Small sample of firefighters. Lack of exposure information.
Douglas et al. 1985 Longitudinal study *[Respiratory-s]*	1,006 London firefighters interviewed and in examined in 1976 and again in 1977.	Years of service, whether fireman had been "punished" by smoke, and if had ever missed a week or more after such an exposure.	Spirometry and prevalence of respiratory symptoms. Average levels of FEV_1, FVC, and FEV_1/FVC were similar to predicted values in both years; all three measures of lung function decreased with age and cigarettes; no association between lung function or respiratory symptoms with smoke exposure or duration of service was found.	Controlled for age and smoking.	Confidence intervals and p-values not reported.
Elci et al. 2003 Hospital-based case-control study *[Cancer-k]*	1,354 male lung cancer patients at a hospital in Turkey, diagnosed and 1,519 male controls diagnosed with other cancer diagnoses (including some non-cancer), 1979-1984.	Employed as firefighter.	Lung cancer: OR 6.8, 95% CI 1.3-37.4.	Age and smoking status adjusted.	Several other occupational also had increased risk of lung cancer (drivers, textile workers, water treatment plant workers, highway construction workers).

continued

TABLE C-1 Continued

Study and Design	Population	Exposure	Outcomes	Adjustments	Limitations and Comments
Eliopulos et al. 1984 Retrospective cohort study [Respiratory-s; Circulatory-k; All Caus?]	990 western Australian firefighters employed between 1939 and 1978. Including 116 deaths.	Employment as a full-time firefighter (yes/no). Duration of employment (years).	SMR and SPMR Cause of death (follow-up through 1978); compared with the male population of Western Australia. All causes: SMR 0.80, 95% CI 0.67-0.96; Other accidents, poisonings, and violence: SMR 0.35, 95% CI 0.10-0.90; All other causes NS. All SPMRs listed NS.	Adjusted for age and calendar year. Stratified by time since first employment and duration of employment; no adjustment for smoking or other potential confounders.	Healthy worker effect.
Elliott et al. 1996 Retrospective cohort study [Cancer-k]	Over 14 million residents living within 7.5 km of a municipal waste incinerator in England (1974-1986), Wales (1974-1984), and Scotland (1975-1987) compared to national cancer incidence rates. Cancer cases identified from a national registry.	Proximity to incinerators (<3 km or 3-7.5 km) at 72 sites in the United Kingdom.	Cancer incidence, relative to general regional population with a 10 year lag (5 year lag for lymphatic and hematopoietic cancers). Stage 1 (22-sites randomly sampled) at 0-3 km All cancers: SIR 1.08, 95% CI 1.07-1.10; Stomach: SIR 1.07, 95% CI 1.02- 1.13; Colorectal: SIR 1.11, 95% CI 1.07- 1.15; Liver: SIR 1.29, 95% CI 1.10- 1.51; Lung: SIR 1.14, 95% CI 1.11 -1.17; Bladder: SIR 1.19, 95% CI 1.13-1.26; Lymphatic and hematopoietic: SIR 1.05; 95% CI 1.01-1.09; non-Hodgkin's lymphoma: SIR 1.11, 95% CI 1.04-1.19. Stage 1 (22-sites randomly sampled) at 0-7.5 km. All cancers: SIR 1.05, 95% CI 1.04-1.05; Stomach: SIR 1.06, 95% CI 1.03-1.09; Colorectal: SIR 1.05, 95% CI 1.03-1.08; Liver: SIR 1.10, 95% CI 1.00- 1.20; Larynx: SIR 1.08, 95% CI 1.00-1.16; Lung: SIR 1.10, 95% CI 1.08-1.12; Bladder: SIR 1.10, 95% CI 1.07-1.14; non-Hodgkin's lymphoma: SIR 1.04, 95% CI 1.01-1.08. Conditional p<0.005 for All cancers, stomach, colorectal, liver, lung and bladder cancers showing significant decline in risk. Stage 2 (remaining 52 sites in the United Kingdom) at 0-3 km. All cancers: SIR 1.04, 95% CI 1.03-1.04; Stomach: SIR 1.05, 95% CI 1.03-1.08); Colorectal: SIR 1.04, 95% CI 1.02-1.06; Liver: SIR 1.13, 95% CI 1.05-1.22; Lung: SIR	Standardized by age, sex, socioeconomic status, region.	Residual confounding may explain the excess of all, stomach, lung and liver cancers near incinerators; the authors expect a substantial level of disease misclassification.

Study	Population	Exposure	Results	Notes
			1.08, 95% CI 1.07-1.09 Stage 2 (remaining 52 sites in the United Kingdom) at 0-7.5 km All cancers: SIR 1.02, 95% 1.02-1.02; Stomach: SIR 1.03, 95% 1.02-1.04; Colorectal: SIR 1.02, 95% 1.01-1.03; Liver: SIR 1.06, 95% 1.01-1.11; Lung: SIR 1.06, 95% 1.05-1.07; Bladder: SIR 1.02, 95% 1.00-1.03. Across both stages, risk for any cancer and stomach, colorectal, liver, and lung cancers decreased significantly with increasing distance from source (both conditional and unconditional p<0.05).	
Feuer and Rosenman 1986 Retrospective cohort study [Respiratory-s; Circulatory-s]	263 white firefighters in New Jersey from 1974-1980 compared to three reference groups: U.S. general population, New Jersey general population, and white police officers.	Employment as a firefighter and in the Police and Fireman Retirement System (10 years of employment or died/became disabled while on the payroll).	PMR Cause of death Compared to U.S. white males: Skin cancer: PMR 2.7; Arteriosclerotic heart disease: PMR 1.22; Bone diseases: PMR 4.00; all significant (p<0.05); All other causes NS. Compared to NJ white males Bone diseases: PMR 3.94 (p<0.05); All other causes NS. Compared to Police Leukemia: PMR 2.76; Respiratory diseases: PMR 1.98; Digestive diseases: PMR 1.54; All significant (p<0.05); All other causes NS. Reference: white policemen Leukemia: 4 (PMR 2.76) Respiratory diseases: 8 (PMR 1.98) Digestive diseases: 25 (PMR 1.54).	Age standardized. Non-white and female cases omitted.
Firth et al. 1996 Retrospective cohort study [Cancer-s]	Cases of cancer among male firefighters, 1972-1984 in the New Zealand Cancer Registry, aged 15-64.	Occupation listed as firefighter on cancer registration.	No significant correlation between duration of employment and any increased mortality. SIR Cancer incidence. Compared to the 1981 New Zealand population; laryngeal cancer among firefighters: SIR 1074, 95% CI 279-2776. No other results reported for firefighters.	Age and socio-economic level standardized. This study only includes cancer cases diagnosed between ages 15 and 64; no information reported for other cancer sites among firefighters.

continued

TABLE C-1 Continued

Study and Design	Population	Exposure	Outcomes	Adjustments	Limitations and Comments
Floret et al. 2003 Population-based case-control study *[Cancers]*	222 cases with non-Hodgkin's lymphoma in Besancon, France, diagnosed 1980-1995 and 2,220 age- and sex-matched controls selected from the 1990 French national census.	Ambient dioxin modeled concentrations at place of residence based on distance from a municipal waste incinerator, categorized into 4 exposure categories (very low, low, intermediate, and high).	Non-Hodgkin's lymphoma Compared to the very low exposure category, Low exposure: OR 1.0, 95% CI 0.7-1.5; Intermediate exposure: OR 0.9, 95% CI 0.6-1.4 High exposure: OR 2.31, 95% CI 1.4-3.8).	Matched for age and sex. Adjusted for socioeconomic status (education, occupation, and household indicators) at the block level.	
Glueck et al. 1996 Prospective cohort study *[Circulatory]*	806 Cincinnati firemen employed between 1984 and 1995. Participants examined every 1 to 4 years. 22 firefighters with CHD.	Employment as a firefighter (yes/no). CHD (yes/no).	Coronary heart disease; myocardial infarction; and risk factors (blood pressure, smoking, fasting glucose, cholesterol); firefighters with incident CHD compared to firefighters without CHD and to healthy employed men participating in the National Health and Nutrition Examination Survey with 8 years of follow-up. Firefighters with CHD had significantly (p<0.05) higher mean age (44 vs. 37; diastolic blood pressure (92 vs. 82 MMHg); systolic blood pressure (140 vs. 125 MMHg); cigarettes per day (12 vs. 3.3); low density lipoprotein cholesterol (148 vs. 127 mg/dl); triglycerides (203 vs. 124 mg/dl); total cholesterol (227 vs. 198 mg/dl); reporting of CHD family history (0.36 vs. 0.17); those without CHD had a longer mean length of follow-up (6.5 vs. 4.3 years). NS differences for mean quetelet score, fasting glucose, high density lipoprotein cholesterol, and length of follow-up. 26 firefighters without CHD and none of the men with CHD suffered severe smoke inhalations.	Adjusted for age, race, and quetelet index (a coronary risk value score).	Few CHD events.
Grimes et al. 1991 Retrospective cohort study	205 deaths among firemen in Honolulu, Hawaii, 1969-1988, compared to the general male population of Hawaii.	Ever employed as firefighter; duration of employment.	RR cause of death Total cohort Genito-urinary cancers: RR 2.28, 95% CI 1.28-4.06; Prostate cancer: RR 2.61, 95% CI 1.38-4.97; Brain and CNS cancer: RR 3.78, 95% CI 1.22-11.71; Circulatory system diseases: RR 1.16, 95% CI 1.10-1.32; Cirrhosis of the liver:	Stratified by race (Caucasian or Hawaiian).	Proportional mortality study. Analysis by person-years not reported. The ethnic composition of the firefighters differs greatly from that of the

153

[Respiratory-s; Neurological-s; Circulatory-s]		RR 2.30, 95% CI 1.21-4.37; All other causes NS. Among Caucasian firefighters: Genito-urinary cancers: RR 3.02, 95% CI 1.49-6.15; Prostate cancer: RR 3.70, 95% CI 1.71-8.02; Brain and CNS cancer: RR 4.15, 95% CI 1.04-16.51; All other causes NS. Among Hawaiian firefighters: Genito-urinary cancers: RR 3.52, 95% CI 1.32-9.36; Prostate cancer: RR 3.35, 95% CI 1.07-10.45; Cirrhosis of the liver: RR 2.99, 95% CI 1.12-7.96; All other causes NS.		general population of Hawaii.
Guidotti et al. 1993 Retrospective cohort study [Respiratory-s; Neurological-s; Circulatory-s; Cancer-k; All Cause]	3,328 male firefighters employed between 1927 and 1987 by the Edmonton or Calgary fire department. 370 deaths.	Years of employment weighted by an exposure opportunity index (relative time exposed to fire by job classification).	SMR Cause of death (1927-1987). Compared with the male population of Alberta, Canada. All malignant neoplasms: SMR 127, 95% CI 102-155; Kidney and ureter cancer: SMR 414, 95% CI 166-853; Mental disorders: SMR 455, 95% CI 274-711; Cerebrovascular: SMR 38.6, 95% CI 17.7-73.3; Digestive system disorders: SMR 46.9, 95% CI 21.4-89.0; External causes: SMR 65.6, 95%CI 48.5-86.7; Suicide: SMR 38.5, 95% CI 15.5-79.2; Other and unknown causes: SMR 31.2, 95% CI 8.5-80.0; All other causes NS. <40 Years of Employment: Lung cancer, cardiovascular disease, obstructive pulmonary disease, and kidney and ureter cancer all NS ≥40 Years of Employment: Kidney and ureter cancer: SMR 3612 (p<0.01), all others NS. Exposure Opportunity Index=0: Obstructive pulmonary disease: SMR 742 (p<0.05); all others NS. Exposure Opportunity Index 0-35: All NS Exposure Opportunity Index ≥35: Lung cancer: SMR 408 (p<0.05); kidney and ureter cancer: SMR 3542 (p<0.01); all others NS.	Age and calendar adjusted; data stratified by exposure opportunity, duration of employment, and cohort of entry (before or after 1950s); latency analysis; no adjustment for smoking or other potential confounders.	Exposure opportunity index not validated empirically or by exposure monitoring; insufficient statistical power to detect lower relative risks and rarer outcomes, or across multiple strata; relies on the accuracy of death certificates; healthy worker effect.

continued

154

TABLE C-1 Continued

Study and Design	Population	Exposure	Outcomes	Adjustments	Limitations and Comments
Gustavsson 1989 Retrospective cohort study [Neurological-s; Circulatory-k; All Cause]	176 Incinerator workers employed for at least 1 year between 1920 and 1985 at a plant near Stockholm, Sweden.	Employment at an incinerator 15 km west of Stockholm, Sweden.	Cause of death (1951-1985); compared with national and local mortality rates. Compared to local mortality rates: All causes NS. Compared to national mortality rates: Lung cancer: SMR 355, 95% CI 162-675; Liver cirrhosis: SMR 454, 95% CI 124-1164; All other causes NS. <40 Years Since First Employment: All causes, All cancers, Lung cancer, Ischemic heart disease NS. ≥40 Years Since First Employment: Ischemic heart disease: SMR 189 ($p<0.01$); All others NS. <30 Years of Employment: All causes, All cancers, Lung cancer, Ischemic heart disease NS. ≥30 Years of Employment: Ischemic heart disease: SMR 167 ($p<0.05$); All others NS	Age and calendar-year adjusted; stratified by length of employment; smoking habits were similar to average Swedish men.	No data on potential confounders and risk factors for disease; possible healthy worker effect.
Gustavsson et al. 1993 Cohort study [Cancer-s]	176 incinerator workers studied from 1951-1985 compared to the male population of Stockholm.	Employed at a waste incinerator.	Mortality from esophageal cancer. SMR 150, 95% CI 4-834.		Follow-up of Gustavsson 1989.
Hansen 1990 Retrospective cohort study [Circulatory-k; All Cause]	886 firefighters out of 48,580 male civil servants and salaried employees, aged between 15 and 69, and employed on the day of the 1970 Denmark census.	Public employees with title of "firefighter" or "fireman" (yes/no).	Cause of death (1970-1980); compared with other civil servants and salaried employees. NS for Malignant neoplasms, lung cancer, other malignant neoplasms, ischemic heart disease, other diseases, and external causes.	Includes analysis by age; no adjustment for smoking or other potential confounders.	Only 10 years of follow up, may not have been sufficient to observe certain effects; healthy worker effect.
Hazucha et al. 2002 Longitudinal study [Respiratory-s]	Non-smoking residents of 3 communities near incinerators and 3 control communities chosen to be similar based on	Residence in a community near an incinerator and daily air quality ($PM_{2.5}$) measurements for each community.	Annual spirometry. There were no significant ($p>0.05$) differences in FVC, FEV_1, FEF_{50} between exposed and control communities, or between the 3 sets of communities.	Community, year, site pair.	Same cohort as Shy et al. 1995, Lee and Shy 1999; Mohan et al. 2000.

Study	Population	Exposure	Outcomes	Adjustments	Comments
Heyer et al. 1990 Retrospective cohort study [Respiratory-s; Circulatory-k; Cancer-s; All Cause]	population density and socioeconomic characteristics over 3 years. 2,289 male firefighters in Seattle, WA, employed for at least 1 year between 1945 and 1980. 383 deaths.	Employment as a firefighter (yes/no); active assignments (having the possibility of fire combat).	Cause of death (1945-1983); compared with U.S. white males. Full cohort All causes: SMR 76, 95% CI 69-85; Circulatory system disease: SMR 78, 95% CI 68-92; Arteriosclerotic disease: SMR 75, 95% CI 63-89; Acute upper airway respiratory disease: SMR 3003, 95% CI 364-10841; Digestive system disease: SMR 43, 95% CI 19-85; Suicide: SMR 21, 95% CI 12-89; All other causes NS. Age <65: All causes: SMR 62, 95% CI 53-71; Circulatory system disease: SMR 68, 95% CI 54-85; Digestive system disease: SMR 37, 95% CI 11-81; All other causes NS. Age 65+: Lung cancer: SMR 177, 95% CI 105-279; All other causes NS. <15 Years of Exposure: All causes: SMR 62, 95% CI 48-79; Circulatory system disease: SMR 63, 95% CI 39-96; All other causes NS. 15-29 Years of Exposure: All causes: SMR 75, 95% CI 65-85; Circulatory system disease: SMR 75, 95% CI 62-91; All other causes NS. 30+ years of exposure: Leukemia: 503, 95% CI 104-1,470; Other lymphatic/ hematopoietic cancer: SMR 989, 95% CI 120-3,571; All other causes NS.	Adjusted for age and calendar year; stratified by duration of exposure and time since first exposure; no adjustment for smoking or other potential confounders.	Reported broad disease categories; included a sub-analysis of those surviving 30 years since first exposure.

continued

155

TABLE C-1 Continued

Study and Design	Population	Exposure	Outcomes	Adjustments	Limitations and Comments
Horsfield et al. 1988a Longitudinal study [Respiratory-s]	96 West Sussex fire brigade firefighters and 69 non-smoking local men as controls followed for 4 years.	Based on employment as firefighter, and self-report of smoke exposure on job.	Self-reported symptoms every 6 months for first 2 years, then yearly (morning coughing, etc). A significant increasing trend in the frequency of reported respiratory symptoms across groups (controls, nonsmoking firemen, ex-smoking firemen, and current smoking firemen), $p<0.0001$ for all symptoms. Symptom frequency increased in firefighters who had previously been exposed to fire smoke compared to unexposed firemen (RR 1.6, $p<0.001$) and remained significant when stratified by smoking (non-smokers: RR 5.7, $p<0.001$; ex-smokers: RR 2.3, $p>0.05$; current smokers: RR 2.3, $p<0.001$). No significant difference in lung function was noted between firefighters exposed to smoke and those not exposed.	Stratified by smoking status.	Selection bias; results indicate a multiplicative effect of smoke exposure and cigarette smoking on reported respiratory symptom frequency.
Horsfield et al. 1988b Longitudinal study [Respiratory-s]	96 West Sussex fire brigade firefighters and 69 non-smoking local men as controls followed for 4 years.	Based on employment as firefighter, and self-report of smoke exposure on job.	Spirometry for lung mechanics and nitrogen washout, every 6 months for first two years, then yearly. Rate of deterioration in lung function greater in controls than in firemen for VC and RV/TLC (both $p<0.05$), FEV_1, FVC, PEF, V_{50}, V_{25}, (all $p<0.01$). No evidence of chronic lung damage found associated with occupation as a fireman.	Results were stratified by smoking; pulmonary function measurements were adjusted for height.	Same cohort as Horsfield et al. 1988; authors note that wearing of breathing equipment may be protective.
Hours et al. 2003 Case-control study [Respiratory-s]	102 male workers from 3 urban incinerators in France compared to 94 male workers from other industries matched on age.	Exposure was categorized by job type: crane or equipment operators (low), furnace workers (medium) and maintenance or effluent treatment workers (high).	Self-reported symptoms; physical exam; red and white blood cell counts, blood lead, liver enzymes; and pulmonary function; incinerator workers reported more symptoms of skin irritation ($p<0.001$), daily cough ($p<0.05$), coughing during specific tasks ($p<0.001$), and had more skin lesions at exam ($p<0.05$); an excess of respiratory problems was also encountered: daily coughing [maintenance and effluent groups (OR 2.55, 95% CI 0.84-7.75); furnace men (OR 6.58, 95% CI 2.18-19.85)]; a significant relationship between exposure and the decrease of several pulmonary parameters was observed between maintenance and effluent workers for expected FEV_1	Regression analyses were adjusted for smoking, age, and work location.	100% participation of workers; exposed workers smoked more cigarettes on average than controls ($p<0.05$); many workers had previous hazardous occupational exposures. Air sampling presented in Maitre et al. 2003.

Study	Population	Exposure	Results	Adjustments	Comments
Hu et al. 2001 Longitudinal study [Respiratory-s]	1018 residents of 3 communities with incinerators or 3 control communities matched on population density and socioeconomic characteristics.	Exposure assessed by community residence (exposed or unexposed), distance from an incinerator, and an incinerator exposure index for 3 months or 12 months of exposure (based on distance and direction from an incinerator).	($p<0.05$), PF ($p<0.001$), and FEV_1/FVC ($p<0.05$) and crane operators for expected FEV_1/FVC ($p<0.05$) and controls. No differences in liver function tests were reported. The incinerator workers had higher levels of neutrophils ($p<0.01$), lymphocyte and monocyte levels (both $p<0.05$), and blood lead levels ($p<0.05$) were elevated in maintenance and effluent workers compared to non-exposed workers. Annual Spirometry, 1992-1994. Overall, there were no significant decreases in pulmonary function between communities with and without incinerators. The 4 different models for each exposure method showed varying decreases (some significant) in percent predicted pulmonary function values for each year but no patterns were observed.	Adjusted for gas oven or range use in the home, length of residence, smoking history.	Exposure may have been too low to elicit any effect over the 3 year time frame.
Jansson and Voog 1989 Case series report/ Ecological study (Register study) [Reproductive-s]	Cases of cleft lip in one Swedish county. Residents 18 Swedish boroughs with incinerators for a Register study.	Exposure was estimated using a dispersion model and residential distance from an incinerator; residence in a borough with an incinerator for the register study.	Incidence of cleft lip and palate; case series: No common explanatory factor was noted for the 6 malformed children born from April to August 1987. All lived from 15-50 km from an incinerator; register study: from 1975-1983, 57 cases of cleft lip or palate were reported (48.5 expected), and 12 cases from 1984-1986 (17.6 expected); no increase in incidence of cleft lip or palate was detected.	None.	No statistical tests (p-values) were reported.
Kang et al. 2008 Registry-based case-control study [Cancer-k]	2,125 white, male firefighter cancer cases in the Massachusetts Cancer Registry compared to 2,763 cancers among white male police officer cancers and 156,890 cancers among all other occupations, 1986-2003.	Primary occupation was firefighter at time of cancer diagnosis.	Cancer incidence Compared to police officers: Colon: SMOR 1.36, 95% CI 1.04-1.79; Brain: SMOR 1.90, 95% CI 1.10-3.26; all other sites NS. All other occupations: No significant findings.	Age and smoking.	Occupational information available for 63% of MCR cases during study period.

157

continued

TABLE C-1 Continued

Study and Design	Population	Exposure	Outcomes	Adjustments	Limitations and Comments
Kilburn et al. 1989 Longitudinal/Case-control study *[Neurological-s]*	14 Firefighters exposed to burning PCBs, assessed 6 months after the fire and again 6-8 weeks after a detoxification program; and 14 non-exposed firefighters.	PCBs and byproducts at a fire event.	Neurophysical and neuropsychological test results. Compared to non-exposed firefighters, exposed firefighters has decreased scores for story memory; visual image; digits backward; block design; trails a and b; identifying embedded figures; design association; recognition; making trails and choice reaction times; and higher POMS for anger, depression, fatigue, and lower for vigor. PCBs in serum and fat were not associated with neurophysical or cognitive impairment. Following a detoxification program, score significantly improved for memory tests, block design, trails B, and embedded figures, and balance among exposed firefighters.	None	
Krishnan et al. 2003 Case-control study *[Cancer-s]*	Cases of glioma in the San Francisco Bay area, 1991-1994 (476 cases, 462 controls) and 1997-1999 (403 cases, 402 controls) and 864 controls matched for age, race, and gender via random digit dialing.	Firefighter occupation (ever/never and as longest held occupation).	For longest held occupation firefighters: OR 5.88, 95% CI 0.70, 49.01. For ever employed as a firefighter: OR 2.85, 95% CI 0.77-10.58. There were no female glioma cases who were employed as firefighters.	Controlled for age, sex, and ethnicity and stratified by time period.	Increased risk of glioma for men employed as janitors and motor vehicle operators. Exposure misclassification- 40% of cases were reported by proxy.
Lee and Shy 1999 Longitudinal study *[Respiratory-s]*	756 Non-smoking residents of 6 communities in southwestern North Carolina, 3 located near an incinerator (<2 miles) and 3 controls communities (>2 miles from an incinerator), 1992-1993.	PM_{10} as a surrogate for outdoor air pollution measured by one station in each community (24-hour mean levels).	Respiratory health diaries and daily measurement of PEFR to determine daily variation; PM_{10} was not related to variations in PEFR nor were there difference between the exposed communities and non-exposed communities; regression of PM_{10} and selected covariates resulted in no association between daily PM_{10} concentrations and PEFR variation however, respiratory hypersensitivity was significantly related to PEFR ($p<0.01$) in both 1992 and 1993; likewise, time spent outdoors in the community was not related; sex, vacuum use, occupational air irritant exposure, age, and height were all related to PEFR ($p<0.05$).	Adjusted for sex, age, respiratory hypersensitivity, hours spent outdoors in the community area, indicators of indoor air pollution (vacuum use and occupational irritants); control communities were selected to have similar socioeconomic characteristics and population density.	

Study	Design	Population	Exposure	Findings	Adjustments	Comments
Liu et al. 1992	Longitudinal study (pre- and post-design) [Respiratory-s]	63 Seasonal and full-time wild land firefighters in Northern California and Montana.	Firefighting for a season in 1989.	Questionnaire, lung function and airway responsiveness tests. Firefighters' post-season results were compared to their pre-season responses and values. Average individual declines in pulmonary function in L/sec: FVC 0.09 (95% CI 0.05-0.13); FEV_1 0.15 (95% CI 0.13-0.17); and FEF_{25-75} 0.44 (95% CI 0.26-0.62). Airway responsiveness significantly increased from pre- to post-season (p=0.02).	Sex, smoking, history of asthma or allergies, years working as a firefighter, respiratory symptoms, membership to a particular firefighting crew.	Small study size.
Lloyd et al. 1988	Ecological study [Reproductive-s]	Single and twin births from 1975-1983 in central Scotland.	Distance from an incinerator.	Rates of twinning in areas exposed to an incinerator compared to non-affected neighboring areas. Only in 1980 was a cluster of twin births detected in exposed areas (p<0.01). There was an anecdotal increase in twinning in cattle during the same time period.	None.	
Ma et al. 1998	Retrospective cohort study [Cancer-s]	6,607 male firefighter deaths (1,883 from cancer) in 24 states, 1984-1993, from the National Occupational Mortality Database.	Usual occupation as a firefighter on death certificate.	Cancer mortality among firefighters, compared to rates of noncancer mortality in same occupational mortality database; among white firefighters. All cancer: MOR 1.11, 95% CI 1.1-1.2; Lip: MOR 5.9, 95% CI 1.9-18.3; Pancreas: MOR 1.2, 95% CI 1.0-1.5; Bronchus and lung: MOR 1.1, 95% CI 1.0-1.2; Soft tissue sarcoma: MOR 1.6, 95% CI 1.0-2.7; Melanoma: MOR 1.4, 95% CI 1.0-1.9; Prostate: MOR 1.2, 95% CI 1.0-1.3; Kidney and renal pelvis: MOR 1.3, 95% CI 1.0-1.7; Non-Hodgkin's lymphoma: MOR 1.4, 95% CI 1.1-1.7; Hodgkin's disease: MOR 2.4, 95% CI 1.4-4.1; All other cancer sites NS. Among black firefighters All cancers: MOR 1.2, 95% CI 0.9-1.5; Brain and central nervous system: MOR 6.9, 95% CI 3.0-16.0; Colon: MOR 2.1, 95% CI 1.1-4.0; Prostate: MOR 1.9, 95% CI 1.2-3.2; Nasopharynx: MOR 7.6, 95% CI 1.3-46.4; All other cancer sites NS.	Age, race, time period adjusted.	Small number of events in some strata; exposure and disease misclassification dependant on the accuracy of death certificates; database similar to Burnett et al. 1994 but extended by 3 years and 3 states not included.

continued

TABLE C-1 Continued

Study and Design	Population	Exposure	Outcomes	Adjustments	Limitations and Comments
Ma et al. 2005 Retrospective cohort study. [Respiratory-s; Neurological-s; Circulatory-s; All Cause]	34,796 male and 2,017 female professional firefighters in Florida from 1972 to 1999. 1,349 male and 38 female deaths.	Certified as a firefighter (yes/no).	Cause of death (1972-1999); compared with general population of Florida. Males All causes: SMR 0.57, 95% CI 0.54-0.60; Infectious diseases: SMR 0.16, 95% CI 0.11-0.22; Allergic/endocrine: SMR 0.35, 95% CI 0.35-0.52; Diabetes: SMR 0.45, 95% CI 0.26-0.73; Mental: SMR 0.41, 95% CI 0.22-0.68; Nervous system: SMR 0.41, 95% CI 0.31-0.86; Circulatory system: SMR 0.69, 95% CI 0.63-76; Cardiovascular: SMR 0.73, 95% CI 0.65-0.83; Respiratory disease: SMR 0.50, 95% CI 0.35-0.70; Pneumonia: SMR 0.34, 95% CI 0.17-0.62; Digestive diseases: SMR 0.57, 95% CI 0.43-0.73; Cirrhosis: SMR 0.49, 95% CI 0.34-0.69; Genitourinary system: SMR 0.8, 95% CI 0.14-0.83; External causes: SMR 0.45, 95% CI 0.40-0.50; All cancers: SMR 0.85, 95% CI 0.77-0.94; Buccal/pharynx cancer: SMR 0.42, 95% CI 0.17-0.87; Pancreatic cancer: SMR 0.57, 95% CI 0.29-0.99; Thyroid cancer: SMR 4.82, 95% CI 1.3-12.3; Breast cancer: SMR 7.41, 95% CI 2.0-19.0; All other causes NS. Females Circulatory system diseases: SMR 2.49, 95% CI 1.32-4.25; Cardiovascular disease: SMR 3.85, 95% CI 1.66-7.58; All other causes NS.	Age, calendar year, and gender adjusted.	Large study size. Disease misclassification dependant on validity of death certificates. Lacks power to detect associations among female firefighters.
Ma et aL 2006 Retrospective cohort study [Cancer-s]	34,796 male and 2,017 female professional firefighters in Florida employed between 1981 and 1999; 970 male and 52 female cancer deaths in the Florida Cancer Data System.	Certification as a firefighter.	Incident cancers among firefighters compared to the general population of Florida. Among male firefighters: All cancer sites: SIR 0.84, 95% CI 0.97-0.90; Buccal: SIR 0.67, 95% CI 0.47-0.91; Digestive: SIR 0.76, 95% CI 0.65-0.89; Stomach: SIR 0.50, 95% CI 0.25-0.90; Respiratory: SIR 0.67, 95% CI 0.57-0.78; Lung/bronchus: SIR 0.65, 95% CI 0.54-0.78; Bladder: SIR 1.29, 95% CI 1.01-1.62; Brain/CNS: SIR 0.58, 95% CI 0.31-0.97; Thyroid: SIR 1.77, 95% CI 1.08-2.73; All	Age and sex adjusted.	Few cancers among female firefighters.

161

Study	Population	Exposure	Results	Adjustments	Comments
			lymphopoietic: SIR 0.68, 95% CI 0.54-0.85; Testes: SIR 1.60, 95% CI 1.20-2.09; All other sites NS. Among female firefighters: All cancer sites: SIR 1.63, 95% CI 1.22-2.14; Thyroid: SIR 3.97, 95% CI 1.45-8.65; Hodgkin's lymphoma: SIR 6.25, 95% CI 1.26-18.30; Cervical: SIR 5.24, 95% CI 2.93-8.65; All other sites NS.		
Markowitz 1989 Longitudinal study [Respiratory-s]	64 exposed firefighters and 22 nonexposed firefighters at Time 2.	Burning PVC at a fire event.	Surveys and pulmonary function tests administered at 5-6 weeks and 22 months after exposure. Firefighters exposed to burning PVC reported more persistent respiratory effects at 6 weeks (cough, wheeze, shortness of breath, and chest pains; each p<0.01) and at 22 months (cough, shortness of breath, and chest pains; each p<0.01) than controls. significanlty more respiratory symptoms at both time points and a physician diagnosis of asthma and/or bronchitis at 22 months for exposed firefighters.	Stratified by smoking and age.	Small study size.
Michelozzi et al. 1998 Ecologic study [Cancer-s]	Residents of Malagrotta, Italy.	Proximity (up to 10 km) to a waste disposal site, a waste incinerator, and an oil refinery operational in the area since the 1960s.	SMR Cause of death (1987-1993) compared to population mortality of the metropolitan Rome area (1991). No significant increased cancer mortality observed in any of the three bands. Among men, mortality from laryngeal cancer declined with distance (p = 0.03), even after adjusting for socioeconomic status (p = 0.06).	Age and sex standardized, adjusted for socioeconomic status.	Pollution may have been from any of the three sources and risks could not be attributable to the incinerator alone.
Miedinger et al. 2007 Cross-sectional study [Respiratory-s]	101 professional firemen and 735 male controls in Basel, Switzerland.	Employment as a firefighter.	Questionnaire, spirometry, skin-prick tests, and bronchial challenge to methacholine. The frequency of reported respiratory symptoms and physician diagnosed asthma did not differ between firefighters and controls but they did report significantly more symptoms while at work (burning eyes, running nose, itchy throat, cough, dyspnea, and headache; all P<0.01). Also elevated rates of sensitization and atopy to several allergens (p<0.001). Increased reactivity to mathancholine was significantly related to being a firefighter (OR 2.25, 95% CI 1.12-4.48); and having an FEV1/ FVC ratio <0.7, wheezing in the last		This study seems to be focused on acute, not chronic, effects. Lack of baseline data for firefighters before exposure.

continued

TABLE C-1 Continued

Study and Design	Population	Exposure	Outcomes	Adjustments	Limitations and Comments
Milham and Ossiander 2001 Cohort study [Circulatory-s]	All deaths among male Washington State residents from 1950-1999. 3145 deaths among firefighters.	Occupation as a firefighter or fire protection worker.	12 months, or doctor diagnosed asthma (all $p<0.01$). Bronchial hyperreactivity was more common among firefighters but could not be linked to acute exposure or duration of employment. Cause of death compared to deaths among other occupations. Significantly ($p<0.05$) decreased PMRs for: Disorders of Character, Behavior & Intelligence (PMR 15); Infective and Parasitic Diseases (PMR 47); Senility & Ill-defined Diseases (PMR 0); AIDS (PMR 28); Tuberculosis of Respiratory System (PMR 15); Tuberculosis (total) (PMR 25); Significantly ($p<0.05$) increased PMRs for: Malignant Melanoma of Skin (PMR 228); Bronchiectasis (PMR 319); MN Lymphatic & Hematopoietic Tissues (PMR 132); Diseases of Other Endocrine Glands (PMR 394); Malignant Neoplasms (total) (PMR 108); Cerebral Hemorrhage and other Vascular Lesions (PMR 115); Lymphosarcoma (PMR 195); All other causes NS.	Age.	Results are able to be queried online at https://fortress.wa.gov/doh/occmort/OMQuery.aspx.
Mohan et al. 2000 Cross-sectional study [Respiratory-s]	4,205 survey respondents from 4 communities in North Carolina and 1 in South Carolina located near 4 incinerators and 4 control communities similar to exposed communities based on socioeconomic and density characteristics.	Residential distance from an incinerator, measured PM10 in each community.	Prevalence of reported respiratory symptoms. ORs comparing each study community to its control community or the combined control group. Significantly fewer ($p<0.05$) Community A respondents reported two or more (OR 0.7), or three or more (OR 0.6) short duration symptoms (eye irritation, sore throat, cough, runny nose, nasal irritation for ≥1 day in the last month) compared with the control community but not against the combined control group; no differences were noted for reporting of long or short duration symptoms; reporting of long or short duration symptoms did not differ between Community B and the control community or combined control group. Community C more frequently reported awakening at night (OR 1.3) vs. the control community but not the combined control	Adjusted for age, gender, ethnicity, length of stay within 2 miles of present residence, education, smoking, use of non-electric source for cooking or heating, central air conditioning, mold in the home, use of humidifier, pets, occupational chemical exposure, perceived air quality.	Analysis combines data previously reported by Shy et al. 1995 and Feigley et al. 1994.

Study	Population	Exposure	Results	Adjustments/Comments
Musk et al. 1982 Prospective cohort study [Respiratory-k]	951 white male firefighters in Boston, MA, studied 1970 to 1976.	Employment as a firefighter (yes/no); calculated and estimated number of fires fought; and by rank or active status.	group; there were no differences for other symptoms; community D residents reported significantly more (p<0.05) wheeze (OR 1.5), morning cough and phlegm or wheeze (OR 1.7), awakening at night (OR 1.8), and two or more (OR 1.6) which remained significant when compared with control community the combined control group; and three or more (OR 1.6) short duration symptoms than the control community but not against the combined control group. Pulmonary function, objective cough, and standardized respiratory symptom questionnaires; at the end of follow-up, FEV_1 was 98.3% of expected; FVC was 97.8% of expected; changes in pulmonary function were not related to firefighting exposure. The authors attribute the results to selection biases and increased use of breathing apparatus in firefighting.	No levels of significance/statistical tests were reported; follow-up of Peters et al. 1974. Analysis considered age, height, and smoking.
Mustacchi 1991 Longitudinal study [Neurological-s]	14 firefighters and 14 controls.	Chemical fires-15-30 minutes of exposure to burning PCBs.	Neurobehavioral functioning measured in 22 tests; exposed firefighters scored significantly differently from controls on profile of mood states (POMS) Total Mood Disturbance Scores; with higher scores for fatigue, and lower scores for vigor (both p<0.05).	Corrected for multiple comparisons.
Mustajbegovic et al. 2001 Cross-sectional study [Respiratory-s]	128 full-time firefighters and 88 factory workers in Zagreb, Croatia.	Employed as a firefighter, duration of employment, and number of fires fought.	Spirometry and prevalence of acute and chronic respiratory symptoms. Dyspnea, nasal catarrh, sinusitis and hoarseness were reported more frequently among firefighters than in control workers (p<0.01); among firefighters, chronic cough, chronic phlegm, chronic bronchitis, and sinusitis were significantly more common among smokers than non-smokers (p<0.01). Increased reporting of chronic phlegm and dyspnea were significantly (p<0.05) associated with length of employment; only chronic bronchitis was associated with increasing age (p<0.05).	Data previously reported by Kilburn et al. 1989. Smoking, age.

163

continued

164

TABLE C-1 Continued

Study and Design	Population	Exposure	Outcomes	Adjustments	Limitations and Comments
			Pulmonary function was significantly decreased among firefighters from expected values for FVC ($p<0.05$), FEV_1 ($p<0.01$), FEF_{50} ($p<0.01$), and FEF75 ($p<0.01$). After controlling for smoking, FEV_1, FEF_{50}, and FEF_{75} remained significantly (all $p<0.005$) related to employment as a firefighter.		
Olshan et al. 1990 Population-based case-control study *[Reproductive-s]*	22,192 live births with birth defects in the British Columbia Health Surveillance registry, 1952-1973; Firefighters fathered 281 live births, 89 with birth defects; police officers fathered 749 live births, 174 with birth defects; among fathers of all other occupations there were 21,929 live births.	Paternal occupation as a firefighter at time of birth surrogate for exposure.	OR Birth defects in 20 categories. All births Compared to other occupations: Ventricular septal defect: OR 2.7, 95% CI 1.02-7.18; Atrial septal defect: OR 5.91, 95% CI 1.60-21.83; All others NS. Compared to Police: Ventricular septal defect: OR 4.68, 95% CI 1.66-13.17; Atrial septal defect: OR 3.76, 95% CI 1.40-10.01; All others NS. Among children with only cardiac defects Compared to other occupations: Atrial septal defects: OR 6.81, 95% CI 1.40-33.16; Ventricular septal defects and Patent ductus arteriosus NS. Compared to Police: Ventricular septal defects: OR 5.05, 95% CI 1.43-17.82; Atrial septal defects: OR 3.82, 95% CI 1.19-12.33; Patent ductus arteriosis: OR 14.60, 95% CI 1.03-206.16.	Matched on date and hospital of birth. Conditional logistic regression for matched controls from all occupations; unconditional logistic regression for comparison with policemen; controlled for father's age and race, and mother's age.	No consideration of maternal exposures.
Peters et al. 1974 Prospective cohort study *[Respiratory-k]*	1,430 Boston firefighters studied 1970-1972.	Employment as a firefighters (yes/no); frequency of fire exposures; number of "shellackings," times overcome at a fire, and frequency of black sputum, and oxygen use.	Percent change in average FVC and FEV_1; self-reported respiratory symptoms over 1 year; FVC and FEV_1 decreased with the number of fires fought ($p<0.01$ and $p<0.02$, respectively); decreases were not able to be explained by age, smoking, or ethnicity; other indicators of exposure (number of shellackings, times overcome at a fire, and frequency of black sputum) were not associated with the decreasing rate of FVC and FEV_1.	Multiple regression analysis adjusting for age, smoking habits, ethnicity.	Assessment only follows firefighters for 1 year.
Prezant et al. 1999	More than 11,000 New York City firefighters	Employment as a firefighter (yes/no). Duration of	Annual incidence and point prevalence of biopsy-proven sarcoidosis (1985-1998); compared with nearly 3,000 EMS workers.	Bivariate correlations between presence of sarcoidosis and length	No exposure free interval.

Retrospective cohort study [Respiratory-s]	employed between 1985 and 1998 and more than 3,000 EMS workers.	employment (years). Company type (engine, ladder, both).	5 cases of biopsy proven sarcoidosis in firefighters and 1 among EMS health care workers; average annual incidence (1995-1998): 12.9 cases/100,000 person-years among firefighters; 0 cases/100,000 person-years among EMS workers; point prevalence on July 1, 1998: 222 cases/100,000 person-years among firefighters; 35 cases/100,000 person-years among EMS workers; no correlations or differences were found for pulmonary function with duration of employment or engine company type for firefighters.	of employment as firefighter and company type.
Rosenstock et al. 1990 Cohort study [Respiratory-s]	4,392 male firefighters (886 deaths) and 2,074 police officers in Seattle, Tacoma, and Portland, OR, employed for at least a year between 1945-1980.	Employment as a firefighter.	Respiratory deaths 1945-1984; compared to U.S. mortality: All cause mortality: SMR 82, 95% CI 11-87; Non-malignant respiratory disease: SMR 81 (95% CI 73-89); malignant respiratory disease and lung, trachea and bronchus cancer NS. Compared to police: All IDR NS.	Adjusted for age and year. Strong healthy worker effect.
Rydhstroem 1998 Ecological study [Reproductive-s]	17,067 twin births in Sweden 1973-1990 from the Medical Birth Registry.	Parental residence in a parish or municipality in the vicinity of one of 14 incinerators.	Incidence of twin deliveries for each year and each municipality; compared to national twinning rates in Sweden; and pre- and post-incinerator implementation twinning rates; by population size: municipalities designated as rural, towns, and other cities were all NS, the 3 largest cities had significantly elevated twinning rates (RR 1.05, 95% CI 1.01-1.10); 6 municipalities (out of 284) had significantly higher twinning rates (p<0.05) in a particular year while two had decreased rates (p<0.05); of the 14 municipalities having an incinerator commissioned between 1973 and 1990, two had significantly reduced (p≤0.05) twinning rates after the plant was commissioned, while one had a higher rate (p<0.01).	Standardized for year of deliver and maternal age.

continued

TABLE C-1 Continued

Study and Design	Population	Exposure	Outcomes	Adjustments	Limitations and Comments
Sama et al. 1990 Registry-based case-control study *[Cancer-e]*	321 cancer cases among firefighters, compared to 392 police and 29, cases of other occupations identified in the Massachusetts Cancer Registry from 1982-1986.	Employed as a firefighter (yes/no).	Cancer incidence of rectal; pancreatic; lung, bronchus and trachea; melanoma of the skin; bladder; brain and other nervous system; non-Hodgkin's lymphoma; and leukemia. Compared to the other occupations in MA: Melanoma of the skin: SMOR 292, 95% CI 170-503; Bladder: SMOR 159, 95% CI 102-250; all others NS. Compared to Police: Bladder: SMOR 211, 95% CI 107-414; Non-Hodgkin's lymphoma: SMOR 327, 95% CI 119-898; all others NS. 18-54 Years of Age compared to Police: Melanoma, Bladder cancer, and Lymphoma all NS. 55-74 Years of Age compared to Police: Melanoma: SMOR 513, 95% CI 150-1750; Lymphoma: SMOR 538, 95% CI 150-1924; Bladder cancer NS. 75+ Years of Age: Melanoma, Bladder cancer, and Lymphoma all NS.	Age standardized.	Occupational information is only available for about half of Massachusetts Cancer Registry registrants.
Sardinas et al. 1986 Retrospective cohort study *[Circulatory-s]*	306 Firefighter and 401 policemen deaths in Connecticut from 1960 to 1978.	Firefighter or policeman listed as "Usual occupation" on death certificates.	115 Ischemic heart disease deaths among firefighters. Comparing firefighters to Connecticut census data, SMR 1.52, 95% CI 1.23-1.81. Comparing firefighters to the standard population, MOR 1.07, 95% CI 0.91-1.23. Comparing firefighters directly to policemen, MOR 0.62, 95% CI 0.56-0.68).	Analysis used weighted averages for 6 time periods.	Only premature deaths between the ages of 25 to 59 were considered.
Schnitzer et al. 1995 Population-based case-control study *[Reproductive-s]*	Cases are infants with major birth defects registered with Atlanta Congenital Defects Program 1968-1980; controls were infants born without defects in same period frequency matched on race, year and hospital of birth.	Parental occupation used as surrogate for occupational exposure to firefighting compared to fathers of other occupations.	Major birth defects Firemen (55 exposed cases) associated with other heart abnormalities (OR 4.7, 95% CI 1.2-17.8), cleft lip (OR 13.3, 95% CI 4.0-44.4), hypospadias (OR 2.6, 95% CI 1.1-6.3), and clubfoot (OR 2.9, 95% CI 1.4-6.0).	Adjusted for mothers alcohol use, smoking, occupation, age, and education; father's smoking and age; and gravidity.	Study identifies several other occupations associated with elevated risks of birth defects.

167

Serra et al. 1996 Cross-sectional study [Respiratory-s]	95 firefighters based at the Sassari Department in Sardinia and a reference group 51 policemen in same province.	Length or service, number of fires fought, pervious work, hobbies, smoking, details about fire stations from questionnaires.	Spirometry, permeability of alveolar-capillary barrier. Significant reduction in some pulmonary function tests among exposed firefighters: FEV_1 ($p<0.05$), FEV_1/FVC ($p<0.001$), FEF_{75} ($p<0.05$), FEF_{50} ($P<0.01$), FEF_{25} ($p<0.001$), and RV ($p<0.01$) but not for FVC, TLC, DLco or DLco/TLC. Lung function was not correlated with years of service, numbers of fires, respiratory data, hobbies or previous professional exposures.	Controlling for age, height, pack-years, and ex-smoker status.	Mostly forest fires, few industrial fires. Limited personal exposure data.
Shy et al. 1995 Cross-sectional study [Respiratory-s]	3475 residents of 3 communities having incinerators and 3402 residents of 3 comparison communities.	Direct measurements and modeled estimates of $PM_{2.5}$ and routinely monitored air pollutants. Residents were members of exposed or unexposed communities; assigned average community exposure measurements; or assigned household exposure estimates determined by residential distance to an incinerator and wind speed/direction.	Pulmonary function, acute or chronic respiratory effects. Normal individuals in exposed and control communities did not have significantly different mean percent predicted FEV_1 and PEFR results; among sensitive individuals, those exposed to a biomedical waste incinerator had significantly lower mean PEFR results compared to the control community and those exposed to a municipal waste incinerator had significantly higher mean FEV_1 and PEFR results than the control community (both $p<0.05$).	Comparison communities had similar socioeconomic characteristics and population densities (subjectively determined by neighborhood appearance) and were no closer than 5 km from the incinerator; adjusted for age, sex, height, and race.	This reports on the first year of a multi-year study.
Sparrow et al. 1992 Longitudinal study [Respiratory-k]	168 firefighters and 1,474 non-firefighters participating in the Normative Aging Study at the VA in Boston, MA.	Employment as a firefighter (yes/no).	Mean annual change in pulmonary function (FVC and FEV_1); compared to non-firefighters; greater loss in FVC and FEV_1 than non-firefighters ($p=0.007$ and $p=0.054$, for FVC and FEV_1, respectively).	Multiple regression models adjusted for smoking status group, age, height, and initial pulmonary function.	Possible healthy worker effects.

continued

TABLE C-1 Continued

Study and Design	Population	Exposure	Outcomes	Adjustments	Limitations and Comments
Tango et al. 2004 Ecological study *[Reproductive-s]*	225,215 live births, 3,387 fetal deaths, and 835 infant deaths among residents living within a 10 km radius around incinerator with dioxin emissions >80 ng/TEQ/m³, 1997-1998, in Japan.	Distance from municipal waste incinerators as a surrogate for exposure; determined from addresses on birth and death certificates.	Congenital malformations, fetal and infant deaths compared to national rates; none of the reproductive outcomes showed statistically significant excess within 2 km of incinerators; a peak-decline in risk from 1-2 km to 10 km was detected for infant deaths (p=0.023) and infant deaths (p=0.047) from all congenital malformations combined.	Adjustment factor to take account of regional differences between study region (all municipalities) and the smaller study area in which study population located.	Lack of detailed exposure information means trend has to be interpreted cautiously.
Tepper et al. 1991 Longitudinal study *[Respiratory-s]*	632 Baltimore firefighters.	Self-reported exposure and fire department records for respiratory use, amonia or chlorine exposure, smoke inhalation, having been overcome by toxic fumes or gases, respiratory symptoms after fires, and months in exposed jobs.	Pulmonary function (FEV_1). Baseline measurements taken 1974-1977 and followed-up 1983-1984; among all firefighters assessed, only work status was associated with significantly decreased FEV_1 such that active firefighters had lower FEV_1 measurements than inactive firefighters; firefighters who ever wore a mask when extinguishing a fire had a significantly smaller annual mean decline in FEV_1 that those that never wore a mask among non-repeater participants but the difference was not significant among all firefighters studied.	Controlled for exposure variables, age, weight, baseline cigarette smoking, and blood type.	Effort to characterize exposure beyond employment; considers issue of respirator use.
Tornling et al. 1994 Retrospective cohort study *[Neurological-s; Circulatory-k; Cancer-k; All Cause]*	1,116 male firefighters in Stockholm, Sweden, employed for at least 1 year between 1931 and 1983. 316 deaths.	Exposure index based on the number of fires fought by each individual.	Cause of death (1958-1986); compared with the greater Stockholm male population. All causes: SMR 82, 95% CI 73-91; Circulatory diseases: SMR 84, 95% CI 71-98; Asthma, bronchitis, and emphysema: SMR 0, 95% CI 0-48; Violent death and poisoning: SMR 52, 95% CI 30-85; Stomach cancer: SMR 192, 95% CI 114-304; All hematopoietic cancers: SMR 32, 95% CI 6-92; All other causes NS. 65+ Years of Age: Brain cancer: SMR 435, 95% CI 140-1015. <30 Years Latency: Ischemic heart disease: SMR 19, 95% CI 4-56; Stomach cancer: SMR 481, 95% CI 155-1122. 30-40 Years Latency: Stomach cancer: SMR 606, 95% CI 313-1059.	Age and calendar year adjusted; stratified by age, employment time, latency period, fire exposure index, and episodes using SCBA; no adjustment for smoking or other potential confounders.	Healthy worker effect; disease misclassification dependant on reliability of death certificate information.

Study	Population	Exposure	Results	Comments
			>30 Years Employment: Stomach cancer: SMR 289, 95% CI 149-505. >1,000 Fires Fought: Stomach cancer: SMR 264, 95% CI 136-461; Brain cancer: SMR 496, 95% CI 1.35-12.70. Other strata by age group, latency, duration of employment, and number of fires all NS for ischemic heart disease, respiratory diseases, stomach cancer, and brain cancer.	
Unger et al. 1980 Cohort study [Respiratory-s]	30 Texan firefighters exposed to severe smoke in a chemical fire outside of Houston.	Severe smoke exposure at particular fire.	Spirometry immediately after exposure, after 6 weeks, and 18 months later. Significant ($p<0.05$) decreases in FVC compared to predicted values immediately after exposure and after 6 weeks, differences were not significant after 18 months; compared to controls, firefighters had significantly decreased FVC ($p<0.01$) and FEV_1 ($p<0.05$) immediately after exposure compared with matched controls.	Healthy controls matched for age, height, and smoking history selected from other laboratory case files. No consideration of disease or exposure prior to the chemical fire or potentially confounding variable such as smoking. Small number of participants.
Vena and Fiedler 1987 Retrospective cohort study. [Respiratory-s; Neurological-s; Circulatory-k; Cancer-k; All Cause]	1,867 white male municipal employees in Buffalo, NY, with at least 5 years of service with at least 1 year as a firefighter between 1950 and 1979. 470 deaths.	Employment as a firefighter (yes/no); length of service (years).	Cause of death (1950-1979); compared with the general U.S. white male population. Colon cancer: SMR 1.83, 95% CI 1.05-2.97; Bladder cancer: SMR 2.86, 95% CI 1.30-5.40; benign neoplasms: SMR 4.17, 95% CI 1.34-9.73; Respiratory diseases: SMR 0.48, 95% CI 0.26-0.80; All external causes: SMR 0.67, 95% CI 0.44-0.98; All other causes NS. 1-9 Years Employed: All causes: SMR 0.59 ($p<0.05$); Circulatory diseases: SMR 0.41 ($p<0.05$); 10-19 Years Employed: All causes: SMR 0.70 ($p<0.05$); 20-29 Years Employed: Brain cancer: SMR 3.75 ($p<0.05$); 30-39 Years Employed: Respiratory diseases: SMR 0.34 ($p<0.05$); 40+ Years Employed: All causes: SMR 1.29 ($p<0.05$); Digestive cancer: SMR 3.08 ($p<0.05$); Colon cancer: SMR 4.71 ($p<0.05$); Bladder cancer: SMR 5.71 ($p<0.05$); All other causes NS.	Stratified by length of service, age, latency, calendar year of death and first employment. No adjustment for smoking or other potential confounders. Potential healthy worker effect.

continued

TABLE C-1 Continued

Study and Design	Population	Exposure	Outcomes	Adjustments	Limitations and Comments
Viel et al. 2000 Retrospective cohort study *[Cancer-s]*	Residents of Doubs, France, 1980-1995.	Proximity to municipal solid waste incinerator.	Soft-tissue sarcoma, non-Hodgkin's lymphoma, Hodgkin's disease. Spatial clusters of increased risk were observed for soft-tissue sarcoma (SIR 1.44, p=0.12) and non-Hodgkin's lymphoma (SIR 1.27, p= 0.0003), but not Hodgkin's disease (SIR 1.42, p=0.95).	Age and sex standardized.	
Viel et al. 2008a Population-based case-control *[Cancer-s]*	2,147 males and 1,827 females with non-Hodgkin's lymphoma in France, 1990-1999, in 2,270 census block groups (cases/controls determined by high/low dioxin exposure).	Exposure to emissions from 13 incinerators was modeled from 1972-1985 (allowing for a 10-year latency period); the exposure model considered incinerator characteristics (capacity, functioning, dust control, fume treatment, operating years), atmospheric diffusion modeling, and distance to estimate the dioxin exposure level of each of the census block groups.	Non-Hodgkin's lymphoma. Highly-exposed block groups relative to lowest exposed: RR 1.12, 95% CI 1.00-1.25. Risks were elevated among women: RR 1.178, 95% CI 1.01-1.37 in the multivariate model.	Population density, urbanization, socio-economic level, airborne traffic pollution, and industrial pollution.	
Vincenti et al. 2008 Ecological study *[Reproductive-s]*	All females aged 16-49 living in the area at any time and females working for at least 1 week from 2003-2006 in 3 municipal areas near Modena, Italy; cases are congenital anomalies and stillbirths identified within 28 days	Residential address and work location history are used as surrogate for exposure and estimated concentration levels. Estimated concentrations are based on estimated fall-out of dioxins and furans in the	Spontaneous abortions, birth defects, still births. No excess risk of miscarriage (RR 1.00, 95% CI 0.65-1.48) or birth defects (RR 0.64, 95% CI 0.20-1.55) in two areas closest to incinerator; also no indication of dose–response trend. No higher risk of spontaneous abortions in females working in factories in exposed areas (RR 1.04, 95% CI 0.38-2.30), but increased prevalence of birth defects (RR 2.26, 95% CI 0.57-6.14).	None.	Likely exposure misclassification. No data on confounders such as smoking, diet, occupation, and reproductive history; small number of cases due to scarcity of exposed women and low number of outcomes.

Study	Population	Exposure	Results	Confounders	Comments
Williams et al. 1992 Ecological study [Reproductive-s]	Residents of several Scottish districts near two incinerators that ceased operation in 3 "at risk" districts and 7 comparison districts were chosen for study.	Districts were determined to be of a priori interest using 3-D mapping techniques based on the probability of exposure incorporating wind direction and strength, influence of topography, anecdotal evidence from residents, and concentrations of pollutants in soil. [lower part of the atmosphere, with 2 areas characterized as having higher levels.] of birth using a regional birth defects registry.	Sex ratios (male:female) for 1975-1979 and 1981-1983. In one "at risk" district (FK4), the m:f ratio was 87, significantly reduced from 100 ($p<0.05$) showing an excess of female births.	None.	Unable to account for potentially confounding variables (parental exposures, etc).
Young et al. 1980 Cross-sectional study [Respiratory-s]	193 New South Wales firefighters.	Length of service (minus time in non-active fire duty) as a firefighter (years).	Questionnaire to determine the prevalence of chronic respiratory symptoms and disease, and pulmonary function testing. Firefighters with chronic bronchitis had a longer average length of exposure ($p<0.05$) and were more likely to be smokers and smoke more (both $p<0.05$) than firefighters without disease but exposure was not related to chronic bronchitis or chronic obstructive airway disease. Changes in pulmonary function were not related to exposure.	Age, smoking, height.	Smoking identified as more significant health risk than fire exposure among this population. This is a pilot study only using 10% of the total cohort.
Zambon et al. 2007 Case-control study [Cancer-s]	172 cases of soft tissue sarcoma diagnosed from 1990-1996 compared with 405 controls matched on age and sex randomly selected from the general population.	Modeled dioxin exposure based on distance between residence and waste incinerators or sources of industrial pollution.	Soft tissue sarcoma. For cases/controls experiencing ≥6 fgr/m³ average exposure compared with <4fgr/m³: OR 2.08, 95% CI 1.19-3.64. For cases/controls experiencing ≥6 fgr/m³ average exposure for ≥32 years compared with <4 fgr/m³: OR 3.30, 95% CI 1.24-8.73. STS risk increased with average exposure among females (p trend = 0.04) but not for men.	Matched on age and sex.	

continued

TABLE C-1 Continued

Study and Design	Population	Exposure	Outcomes	Adjustments	Limitations and Comments
Studies of Gulf War veterans					
Barth et al. 2009 Cohort study	621,902 U.S. GW veterans and 746,248 nondeployed era veterans followed through 2004.	Estimated oil-well fire smoke modeled.	123,478 GW Veterans were exposed to oil-well fire smoke. Increased risk of brain cancer among veterans exposed to oil-well fire smoke compared to non-exposed veterans (OR 1.67, 95% CI 1.05-2.65) and after controlling for 2+ days of exposure to nerve agents at Khamisiyah (OR 1.81, 95% CI 1.00-3.27).	Controlling for sex, race, age, and unit type.	Follow-up of Kang and Bullman 2001.
Bullman et al. 2005 Cohort study	100,487 U.S. Army GW veterans exposed to chemical warfare agents at Khamisiyah; 224,980 unexposed Army GW veterans; exposure determined from the DoD plume model.	Exposure to oil-well fires and nerve agents determined by plume model.	Brain cancer mortality Oil-well fire smoke was not significantly related to brain cancer deaths.	Controlling for sex, race, age, and unit type.	Follow-up of Kang and Bullman 2001.
Cowan et al. 2002 Case-control study	873 cases of asthma compared to 2464 controls using a DoD registry, among GW veterans.	Modeled oil-well fire smoke.	Physician-diagnosis of asthma 3-6 years after war. Asthma associated with cumulative exposure (<0.1 mg/m^3/day referent) between 0.1-1.0 mg/m^3/day (OR 1.24, 95% CI 1.00-1.55); for exposure of 1mg/m^3/day or greater (OR 1.40, 95% CI 1.11-1.75); and as a continuous variable, OR 1.08, 95% CI 1.01-1.15. Number of days at >65 µg/m^3 compared to 0 days of exposure: 1-5 days of exposure (OR 1.22, 95% CI 0.99-1.51; 6-30 days of exposure (OR 1.41, 95% CI 1.12-1.77); and as a continuous variable (OR 1.03, 95% CI 1.01-1.05).	Sex, age, race, military rank, smoking history, self-reported exposure.	Pre-exposure asthma status of participants unknown.
Iowa Persian Gulf Study Group 1997 Population-based cross-sectional study	3,695 GW veterans and non-GW veterans living in Iowa.	Oil-well fire smoke exposure collected by questionnaire.	Prevalence of self-reported symptoms and illnesses. Many reported exposures were significantly related to multiple self-reported symptoms or illnesses; 85.2% of regular military and 96% of National Guard/Reserve GW veterans reported	Stratified for age, sex, rank, race, and branch of service.	Study not designed to investigate the effects of exposure to oil-well fire smoke.

Study	Population	Exposure	Results	Adjustments	Comments
Kang et al. 2000 Cross-sectional study	15,000 GW veterans and 15,000 non-GW veterans, selected by stratified random sample.	Exposure to oil-well fires assessed by questionnaire.	exposure to smoke or combustion products. Exposure to smoke/combustion products was associated with depression, cognitive dysfunction, and fibromyalgia (all p<0.001). Prevalence of selected self-reported medical conditions. 65% of GW veterans, and 73% veterans in the VA Health Registry reported exposure to oil-well fire smoke but no significant differences in prevalence of self-reported medical conditions was reported.	Stratified by sex and component.	Study not designed to investigate the effects of exposure to oil-well fire smoke.
Lange et al. 2002 Cross-sectional study	1,560 GW veterans.	Self-reported and modeled exposure to oil-well fire smoke.	Asthma and bronchitis symptoms collected by structured interviews conducted 5 years after the war. No association between modeled exposure and asthma or bronchitis symptoms. Self-reported exposure >30 days was significantly related to asthma (OR 2.83) and bronchitis (OR 4.78) symptoms.	Sex, age, race, military rank, smoking history, military service, level of preparedness for war.	Used symptom-based case definition of bronchitis and asthma (possible disease misclassification).
Proctor et al. 1998 Longitudinal study	Stratified random sample of 220 GW veterans from Ft. Devens and 71 from New Orleans, and 50 Era veterans deployed to Germany, assessed 1994-1996.	Combustion products.	Smoke from oil well fires was not significantly related to any system diseases but smoke from burning human waste was related to cardiac symptoms (p<0.001) and pulmonary symptoms (p<0.015).	Controlled for age, sex, education, study site, PTSD status, and war-zone exposure.	Vehicle exhaust related to cardiac and neurological symptoms; and smoke from tent heaters related to cardiac, neurological and pulmonary outcomes.
Smith et al. 2002 Cohort study	405,142 active-duty GW veterans who were in theater during the time of Kuwaiti oil-well fires.	Modeled PM exposure to represent oil-well fire smoke exposure used to create 7 categories of exposure. 1: average daily exposure of 1-260 µg/m3 for 1-25 days; Exposure level 2: average daily exposure of 1-260 µg/m3 for 26-50 days; Exposure level 3: average daily	DoD hospitalizations 1991-1999. Hospitalization rates among those in exposure groups 1-6 were compared to personnel determined to be unexposed. Only exposure level 4 was at increased risk of hospitalization (RR 1.03, 95% CI 1.00-1.05), risks for all other exposure levels NS. Causes for hospitalization and levels of exposure: Infections and parasitic diseases: level 2 (RR 0.87, p<0.05); Endocrine, nutritional, and metabolic disorders: level 1 (RR 0.89, p<0.05); level 4 (RR 0.87, p<0.05); level 5 (RR 0.84, p<0.05); level 6 (RR 0.84, p<0.05); Mental disorders: level 6 (RR 1.11, p<0.05);	Adjusted for "influential covariates," defined as demographic or deployment variables with p values less than 0.15.	Objective measure of disease not subject to recall bias; no issues with self-selection; however, only DoD hospitals, only active duty, no adjustment for potential confounders such as smoking.

continued

TABLE C-1 Continued

Study and Design	Population	Exposure	Outcomes	Adjustments	Limitations and Comments
		exposure of 1-260 µg/m3 for >50 days; Exposure level 4: average daily exposure of >260 µg/m3 for 1-25 days; Exposure level 5: average daily exposure of >260 µg/m3 for 26-50 days. Exposure level 6: average daily exposure of >260 µg/m3 for >50 days.	Circulatory diseases: level 2 (RR 0.88, p<0.05); level 4 (RR 0.87, p<0.05); level 5 (RR 0.90, p<0.05); Respiratory diseases: level 3 (RR 0.69, p<0.05); Digestive diseases: level 5 (RR 0.92, p<0.05); Genitourinary diseases: level 5 (RR 0.91, p<0.05); Pregnancy complications: level 2 (RR 0.86, p<0.05); level 3 (RR 0.48, p<0.05); level 6 (0.84, p<0.05); Skin diseases: level 3 (RR 1.35, p<0.05); level 5 (RR 0.87, p<0.05); Musculoskeletal disease: level 2 (RR 0.91, p<0.05); Symptoms, signs, and ill0defined conditions: level 4 (RR 0.92, p<0.05); level 5 (RR 0.90, p<0.05); Injury and poisoning: level 4 (RR 1.11, p<0.05); All other causes and levels of exposure NS.		
Spencer et al. 2001 Case-control study	241 veterans meeting the criteria for unexplained illness in Washington or Oregon and 113 health veterans as controls.	GW combat (heat stress, chemical exposures, oil-well fire smoke).	Prevalence of unexplained illness (by PEHRC or CDC definitions) assessed by survey and clinical study; burned latrine waste exposure was associated with unexplained illness (OR 2.51, 95% CI 1.58-3.98); many exposures were significantly related to unexplained illness, most strongly being sun exposure, conditions of combat, and medical problems/treatment sought while deployed.	Controlled for other simultaneous exposures.	
Unwin et al. 1999 Cross-sectional study	8,195 GW veterans and Bosnia and other era veterans deployed elsewhere from the United Kingdom, conducted in 1997-1998.	Combustion product exposure assessed by questionnaire.	Prevalence of self-reported symptoms and illnesses. Among all three groups of veterans, exposure to oil-well fire smoke was not significantly associated with physical functioning. For CDC syndrome, risks were increased among GW veterans (OR 1.8, 95% CI 1.5-2.1) and era veterans (OR 1.8, 95% CI 1.1-2.9). Risks were increased for PTSD among GW veterans (OR 2.3, 95% CI 1.7-2.9), Bosnia veterans (OR 3.2, 95% CI 1.6-6.8) and era veterans (OR 3.0, 95% CI 1.4-6.5).	Stratified for age, rank, and deployment to Bosnia.	

Verret et al. 2008 Cross sectional and nested case-control study	5,666 French GW veterans.	Oil-well fire.	Fertility disorders, miscarriage, birth defects assessed by questionnaire in 2002-2004. 0.9% reported infertility, 12% reported one or more miscarriages among partners of male veterans, 2.4% fathered children with a birth defect conceived after returning from the Persian Gulf. Case-control study comparing exposures experienced by fathers with and without children having birth defects- No exposure (time of mission, location of mission, oil-well fire smoke, sandstorm, chemical arms, and pesticides) was related to birth defects. Incidence of birth defects among veterans was similar to that of the French population (except for Down syndrome, RR 0.36, 95% CI 0.13-0.78).	Adjusted for age, service branch, rank, and military status. Controls to compare the risk of birth defects with specific exposures were male veterans who never had a child with a birth defect but had at least one healthy child matched with veterans who fathered a child with a birth defect after deployment on age.
White et al. 2001 Cross-sectional study	193 GW veterans and 47 Germany deployed veterans.	Deployment to the GW 1990-1991 and related self-reported exposures.	Neuropsychological function. Chemical warfare agent and pesticide exposures were related to poorer neuropsychological tests performance (p<0.05), oil well fire smoke, pyridostigmine bromide were not. Exposure to oil well fire smoke significantly increased score on the POMS tension scale; GW veterans performed worse on several tests (only mood complaints remained significant after Bonferroni correction) than Germany-deployed veterans.	Adjusted for age, education, gender, and sampling design. Controlled for post-traumatic stress disorder, major depression, and other known covariates. Adjustment for multiple comparisons.
Wolfe et al. 2002 Cross-sectional study	1,290 GW veterans at Ft Devens in 1997 (who previously were surveyed in 1991).	Deployment to the GW 1990-1991 and related exposures.	Prevalence of multisymptom illness (at least two categories of symptoms: fatigue, mood-cognition, musculoskeletal). 60% prevalence of multisymptom illness. Multivariate regression showed several factors to be related (female, OR 1.8, 95% CI 1.1-2.9; college education, OR 0.5, 95% CI 0.4-0.7; GSI clinical caseness, OR 9.8, 95% CI 7.3-13.1; oil fire smoke, OR 1.6, 95% CI 1.2-2.1; chemicals, OR 2.4, 95% CI 1.6-3.6; heater in tent, OR 1.4, 95% CI 1.0-1.8; seen in clinic, OR 1.5, 95% CI 1.2-2.0; anthrax vaccine, OR 1.5, 95% CI 1.1-2.0; medium or exposure to Anti-nerve gas, OR 1.4, 95% CI 1.0-1.9 and OR 2.1, 95% CI 1.4-3.1 respectively.	Stratified for GSI caseness criteria.

TABLE C-1 Continued

NOTE: AIDS = acquired immune deficiency syndrome; CDC = Centers for Disease Control and Prevention; CHD = coronary heart disease; CI = confidence interval; CNS = central nervous system; DoD = Department of Defense; EMS = emergency medical services; FEF_{25-75} = forced expiratory flow between 25% and 75%; FEF_{50} = forced expiratory flow at 50%; FEV_1 = forced expiratory volume in one second; FVC = forced vital capacity; GW = Gulf War; IDR = incidence density ratio; IRR = incidence rate ratio; km = kilometers; L/sec = liters per second; MN = malignant; MOR = mortality odds ratio; NS = not significant; OR = odds ratio; PEFR = peak expiratory flow rate; PM = particulate matter; PMR = proportional mortality ratio; PVC = polyvinyl chloride; RR = relative risk; RV = residual volume; SIR = standardized incidence ratio; SMR = standardized mortality ratio; SMOR = standardized mortality odds ratio; SPMR = standardized proportional mortality ratio; TEQ = toxicity equivalent; TLC = total lung capacity; V_{25} = maximum expiratory flow rates at 25% of FVC; V_{50} = maximum expiratory flow rates at 50% of FVC; VA = Department of Veterans Affairs; VC = vital capacity.

REFERENCES

Aronson, K. J., L. A. Dodds, L. Marrett, and C. Wall. 1996. Congenital anomalies among the offspring of fire fighters. *American Journal of Industrial Medicine* 30(1):83-86.

Aronson, K. J., G. A. Tomlinson, and L. Smith. 1994. Mortality among fire fighters in metropolitan Toronto. *American Journal of Industrial Medicine* 2(1):89-101.

Bandaranayke, D., D. Read, and Clare Salmond. 1993. Health consequences of a chemical fire. *International Journal of Environmental Health Research* (3):104-114.

Baris, D., T. J. Garrity, J. L. Telles, E. F. Heineman, A. Olshan, and S. H. Zahm. 2001. Cohort mortality study of Philadelphia firefighters. *American Journal of Industrial Medicine* 39(5):463-476.

Barth, Shannon K., Han K. Kang, Tim A. Bullman, and Mitchell T. Wallin. 2009. Neurological mortality among U.S. veterans of the Persian Gulf War: 13-year follow-up. *American Journal of Industrial Medicine* 52(9):663-670.

Bates, J. T. 1987. Coronary artery disease deaths in the Toronto Fire Department. *Journal of Occupational Medicine* 29(2):132-135.

Bates, M. N. 2007. Registry-based case-control study of cancer in California firefighters. *American Journal of Industrial Medicine* 50(5):339-344.

Bates, M. N., J. Fawcett, N. Garrett, R. Arnold, N. Pearce, and A. Woodward. 2001. Is testicular cancer an occupational disease of fire fighters? *American Journal of Industrial Medicine* 40(3):263-270.

Beaumont, J. J., G. S. Chu, J. R. Jones, M. B. Schenker, J. A. Singleton, L. G. Piantanida, and M. Reiterman. 1991. An epidemiologic study of cancer and other causes of mortality in San Francisco firefighters. *American Journal of Industrial Medicine* 19(3):357-372.

Betchley, C., J. Q. Koenig, G. van Belle, H. Checkoway, and T. Reinhardt. 1997. Pulmonary function and respiratory symptoms in forest firefighters. *American Journal of Industrial Medicine* 31(5):503-509.

Biggeri, A., F. Barbone, C. Lagazio, M. Bovenzi, and G. Stanta. 1996. Air pollution and lung cancer in Trieste, Italy: Spatial analysis of risk as a function of distance from sources. *Environmental Health Perspectives* 104(7):750-754.

Bresnitz, E. A., J. Roseman, D. Becker, and E. Gracely. 1992. Morbidity among municipal waste incinerator workers. *American Journal of Industrial Medicine* 2(3):363-378.

Bullman, T. A., C. M. Mahan, H. K. Kang, and W. F. Page. 2005. Mortality in US Army Gulf War veterans exposed to 1991 Khamisiyah chemical munitions destruction. *American Journal of Public Health* 95(8):1382-1388.

Burnett, C. A., W. E. Halperin, N. R. Lalich, and J. P. Sestito. 1994. Mortality among fire fighters: a 27 state survey. *American Journal of Industrial Medicine* 26(6):831-833.

Calvert, G. M., J. W. Merling, and C. A. Burnett. 1999. Ischemic heart disease mortality and occupation among 16- to 60-year-old males. *Journal of Occupational Medicine* 41(11):960-966.

Carozza, S. E., M. Wrensch, R. Miike, B. Newman, A. F. Olshan, D. A. Savitz, M. Yost, and M. Lee. 2000. Occupation and adult gliomas. *Am J Epidemiol* 152(9):838-846.

Charbotel, B., M. Hours, A. Perdrix, L. Anzivino-Viricel, and A. Bergeret. 2005. Respiratory function among waste incinerator workers. *International Archives of Occupational and Environmental Health* 78(1):65-70.

Comba, P., V. Ascoli, S. Belli, M. Benedetti, L. Gatti, P. Ricci, and A. Tieghi. 2003. Risk of soft tissue sarcomas and residence in the neighbourhood of an incinerator of industrial wastes. *Occupational and Environmental Medicine* 60(9):680-683.

Cordier, S., C. Chevrier, E. Robert-Gnansia, C. Lorente, P. Brula, and M. Hours. 2004. Risk of congenital anomalies in the vicinity of municipal solid waste incinerators. *Occupational and Environmental Medicine* 61(1):8-15.

Cowan, DN, Lange JL, Heller J, Kirkpatrick J, DeBakey S. 2002. A case-control study of asthma among U.S. Army Gulf War veterans and modeled exposure to oil well fire smoke. *Military Medicine* 167(9):777-782.

Cresswell, P. A., J. E. Scott, S. Pattenden, and M. Vrijheid. 2003. Risk of congenital anomalies near the Byker waste combustion plant. *Journal of Public Health Medicine* 2 (3):237-242.

Demers, P. A., H. Checkoway, T. L. Vaughan, N. S. Weiss, N. J. Heyer, and L. Rosenstock. 1994. Cancer incidence among firefighters in Seattle and Tacoma, Washington (United States). *Cancer Causes Control* 5(2):129-135.

Demers, P. A., N. J. Heyer, and L. Rosenstock. 1992a. Mortality among firefighters from three northwestern United States cities. *British Journal of Industrial Medicine* 49(9):664-670.

Demers, P. A., T. L. Vaughan, H. Checkoway, N. S. Weiss, N. J. Heyer, and L. Rosenstock. 1992b. Cancer identification using a tumor registry versus death certificates in occupational cohort studies in the United States. *American Journal of Epidemiology* 136(10):1232-1240.

Deschamps, S., I. Momas, and B. Festy. 1995. Mortality amongst Paris fire-fighters. *European Journal of Epidemiology* 11(6):643-646.

Dibbs, E., H. E. Thomas, S. T. Weiss, and D. Sparrow. 1982. Fire fighting and coronary heart disease. *Circulation* 65 (5):943-946.

Douglas, D. B., R. B. Douglas, D. Oakes, and G. Scott. 1985. Pulmonary function of London firemen. *British Journal of Industrial Medicine* 42(1):55-58.

Elci, O. C., M. Akpinar-Elci, M. Alavanja, and M. Dosemeci. 2003. Occupation and the risk of lung cancer by histologic types and morphologic distribution: a case control study in Turkey. *Monaldi Archives of Chest Disease* 59(3):183-188.

Eliopulos, E., B. K. Armstrong, J. T. Spickett, and F. Heyworth. 1984. Mortality of fire fighters in Western Australia. *British Journal of Industrial Medicine* 41(2):183-187.

Elliott, P., G. Shaddick, I. Kleinschmidt, D. Jolley, P. Walls, J. Beresford, and C. Grundy. 1996. Cancer incidence near municipal solid waste incinerators in Great Britain. *British Journal of Cancer* 73(5):702-710.

Feuer, E., and K. Rosenman. 1986. Mortality in police and firefighters in New Jersey. *American Journal of Industrial Medicine* 9(6):517-27.

Firth, H. M., K. R. Cooke, and G. P. Herbison. 1996. Male cancer incidence by occupation: New Zealand, 1972-1984. *International Journal of Epidemiology* 25(1):14-21.

Floret, N., F. Mauny, B. Challier, P. Arveux, J. Y. Cahn, and J. F. Viel. 2003. Dioxin emissions from a solid waste incinerator and risk of non-Hodgkin lymphoma. *Epidemiology* 14(4):392-398.

Glueck, CJ, et al. 1996. Risk Factors for Coronary Heart Disease Among Firefighters in Cincinnati. *American Journal of Industrial Medicine* 30:331-340.

Grimes, G., D. Hirsch, and D. Borgeson. 1991. Risk of death among Honolulu fire fighters. *Hawaii Medical Journal* 50(3):82-85.

Guidotti, T. L. 1993. Mortality of urban firefighters in Alberta, 1927-1987. *American Journal of Industrial Medicine* 23(6):921-940.

Gustavsson, P. 1989. Mortality among workers at a municipal waste incinerator. *American Journal of Industrial Medicine* 15(3):245-253.

Gustavsson, P., B. Evanoff, and C. Hogstedt. 1993. Increased risk of esophageal cancer among workers exposed to combustion products. *Archives of Environmental Health* 48(4):243-245.

Hansen, E. S. 1990. A cohort study on the mortality of firefighters. *British Journal of Industrial Medicine* 47(12):805-809.

Hazucha, M. J., V. Rhodes, B. A. Boehlecke, K. Southwick, D. Degnan, and C. M. Shy. 2002. Characterization of spirometric function in residents of three comparison communities and of three communities located near waste incinerators in North Carolina. *Archives of Environmental Health* 5(2):103-112.

Heyer, N., N. S. Weiss, P. Demers, and L. Rosenstock. 1990. Cohort mortality study of Seattle fire fighters: 1945-1983. *American Journal of Industrial Medicine* 17(4):493-504.

Horsfield, K., F. M. Cooper, M. P. Buckman, A. R. Guyatt, and G. Cumming. 1988a. Respiratory symptoms in West Sussex firemen. *British Journal of Industrial Medicine* 45(4):251-255.

Horsfield, K., A. R. Guyatt, F. M. Cooper, M. P. Buckman, and G. Cumming. 1988b. Lung function in West Sussex firemen: a four year study. *British Journal of Industrial Medicine* 45(2):116-121.

Hours, M., L. Anzivino-Viricel, A. Maitre, A. Perdrix, Y. Perrodin, B. Charbotel, and A. Bergeret. 2003. Morbidity among municipal waste incinerator workers: A cross-sectional study. *International Archives of Occupational and Environmental Health* 76(6):467-472.

Hu, S. W., M. Hazucha, and C. M. Shy. 2001. Waste incineration and pulmonary function: an epidemiologic study of six communities. *Journal of the Air & Waste Management Association* 51(8):1185-1194.

Iowa Persian Gulf Study Group. 1997. Self-reported illness and health status among Gulf War veterans: A population-based study. *Journal of the American Medical Association* 27(3):238-245.

Jansson, B., and L. Voog. 1989. Dioxin from Swedish municipal incinerators and the occurrence of cleft lip and palate malformations. *International Journal of Environmental Studies* (34):99-104.

Kang, D., L. K. Davis, P. Hunt, and D. Kriebel. 2008. Cancer incidence among male Massachusetts firefighters, 1987-2003. *American Journal of Industrial Medicine* 51(5):329-335.

Kang, H. K., C. M. Mahan, K. Y. Lee, C. A. Magee, and F. M. Murphy. 2000. Illnesses among United States veterans of the Gulf War: a population-based survey of 30,000 veterans. *Journal of Occupational and Environmental Medicine* 42(5):491-501.

Kilburn, K. H., R. H. Warsaw, and M. G. Shields. 1989. Neurobehavioral dysfunction in firemen exposed to polycholorinated biphenyls (PCBs): possible improvement after detoxification. *Archives of Environmental Health* 44(6):345-350.

Krishnan, G., M. Felini, S. E. Carozza, R. Miike, T. Chew, and M. Wrensch. 2003. Occupation and adult gliomas in the San Francisco Bay Area. *Journal of Occupational and Environmental Medicine* 45(6):639-647.

Lange, J.L., D. A. Schwartz, B. N. Doebbeling, J. M. Heller, and P. S. Thorne. 2002. Exposures to the Kuwait oil fires and their association with asthma and bronchitis among gulf war veterans. *Environmental Health Perspectives* 110(11):1141-1146.

Lee, J. T., and C. M. Shy. 1999. Respiratory function as measured by peak expiratory flow rate and PM10: six communities study. *Journal of Exposure Analysis and Environmental Epidemiology* 9(4):293-299.

Liu, D., I. B. Tager, J. R. Balmes, and R. J. Harrison. 1992. The effect of smoke inhalation on lung function and airway responsiveness in wildland fire fighters. *American Review of Respiratory Disease* 146(6):1469-1473.

Lloyd, O. L., M. M. Lloyd, F. L. Williams, and A. Lawson. 1988. Twinning in human populations and in cattle exposed to air pollution from incinerators. *British Journal of Industrial Medicine* 4(8):556-560.

Ma, F., L. E. Fleming, D. J. Lee, E. Trapido, and T. A. Gerace. 2006. Cancer incidence in Florida professional firefighters, 1981 to 1999. *Journal of Occupational and Environmental Medicine* 48(9):883-888.

Ma, F., L. E. Fleming, D. J. Lee, E. Trapido, T. A. Gerace, H. Lai, and S. Lai. 2005. Mortality in Florida professional firefighters, 1972 to 1999. *American Journal of Industrial Medicine* 47(6):509-517.

Ma, F., D. J. Lee, et al. 1998. Race-specific cancer mortality in US firefighters: 1984-1993. *Journal of Occupational and Environmental Medicine* 40(12):1134-1138.

Markowitz, J. S. 1989. Self-reported short- and long-term respiratory effects among PVC-exposed firefighters. *Archives of Environmental Health* 44(1):30-33.

Michelozzi, P., F. Forastiere, D. Fusco, C. A. Perucci, B. Ostro, C. Ancona, and G. Pallotti. 1998. Air pollution and daily mortality in Rome, Italy. *Occupational and Environmental Medicine* 55(9):605-610.

Miedinger, D., P. N. Chhajed, D. Stolz, C. Gysin, A. B. Wanzenried, C. Schindler, C. Surber, H. C. Bucher, M. Tamm, and J. D. Leuppi. 2007. Respiratory symptoms, atopy and bronchial hyperreactivity in professional firefighters. *European Respiratory Journal* 30(3):538-544.

Milham, S., and E. Ossiander. 2011. *Occupational mortality database*. Washington State Department of Health. https://www.doh.wa.gov/data/multi-topic.htm (accessed January 19, 2011).

Mohan, A., D. Degnan, C.E. Feigley, C. M. Shy, C.A. Hornung, T. Mustafa, and C.A. Macera. 2000. Comparison of respiratory symptoms among community residents near waste disposal incinerators. *International Journal of Environmental Health Research* 10:63-75.

Musk, A. W., J. M. Peters, L. Bernstein, C. Rubin, and C. B. Monroe. 1982. Pulmonary function in firefighters: a six-year follow-up in the Boston Fire Department. *American Journal of Industrial Medicine* 3(1):3-9.

Mustacchi, P. 1991. Neurobehavioral dysfunction in firemen exposed to polychlorinated biphenyls (PCBs): possible improvement after detoxification. *Archives of Environmental Health* 46(4):254-255.

Mustajbegovic, J., E. Zuskin, E. N. Schachter, J. Kern, M. Vrcic-Keglevic, S. Heimer, K. Vitale, and T. Nada. 2001. Respiratory function in active firefighters. *American Journal of Industrial Medicine* 40(1):55-62.

Olshan, A. F., K. Teschke, and P. A. Baird. 1990. Birth defects among offspring of firemen. *Am J Epidemiol* 131 (2):312-21.

Peters, J. M., G. P. Theriault, L. J. Fine, and D. H. Wegman. 1974. Chronic effect of fire fighting on pulmonary function. *New England Journal of Medicine* 291(25):1320-1322.

Prezant, D. J., A. Dhala, A. Goldstein, D. Janus, F. Ortiz, T. K. Aldrich, and K. J. Kelly. 1999. The incidence, prevalence, and severity of sarcoidosis in New York City firefighters. *Chest* 116(5):1183-1193.

Proctor S.P., T. Heeren, R. F. White, J. Wolfe, M. S. Borgos, J. D. Davis, L. Pepper, R. Clapp, P. B. Sutker, J. J. Vasterling, and D. Ozonoff. 1998. Health status of Persian Gulf War veterans: Self-reported symptoms, environmental exposures and the effect of stress. *International Journal of Epidemiology* 27(6):1000-1010.

Rosenstock, L., P. Demers, N. J. Heyer, and S. Barnhart. 1990. Respiratory mortality among firefighters. *British Journal of Industrial Medicine* 47(7):462-465.

Rydhstroem, H. 1998. No obvious spatial clustering of twin births in Sweden between 1973 and 1990. *Environmental Research* 76(1):27-31.

Sama, S. R., T. R. Martin, L. K. Davis, and D. Kriebel. 1990. Cancer incidence among Massachusetts firefighters, 1982-1986. *American Journal of Industrial Medicine* 18(1):47-54.

Sardinas, A., J. W. Miller, and H. Hansen. 1986. Ischemic heart disease mortality of firemen and policemen. *American Journal of Public Health* 76(9):1140-1141.

Schnitzer, P. G., A. F. Olshan, and J. D. Erickson. 1995. Paternal occupation and risk of birth defects in offspring. *Epidemiology* 6(6):577-583.

Serra, A., F. Mocci, and F. S. Randaccio. 1996. Pulmonary function in Sardinian fire fighters. *American Journal of Industrial Medicine* 30(1):78-82.

Shy, C. M., D. Degnan, D. I. Fox, S. Mukerjee, M. J. Hazucha, B. A. Boehlecke, D. Rothenbacher, P. M. Briggs, R. B. Devlin, D. D. Wallace, R. K. Stevens, and P. A. Bromberg. 1995. Do waste incinerators induce adverse respiratory effects? An air quality and epidemiological study of six communities. *Environmental Health Perspectives* 103(7-8):714-724.

Smith, T. C., J. M. Heller, T. I. Hooper, G. D. Gackstetter, and G. C. Gray. 2002. Are Gulf War veterans experiencing illness due to exposure to smoke from Kuwaiti oil well fires? Examination of Department of Defense hospitalization data. *American Journal of Epidemiology* 155(10):908-917.

Sparrow, D., R. Bosse, et al. 1982. The effect of occupational exposure on pulmonary function: a longitudinal evaluation of fire fighters and nonfire fighters. *American Review of Respiratory Disease* 125(3):319-322.

Spencer, P. S., L. A. McCauley, J. A. Lapidus, M. Lasarev, S. K. Joos, and D. Storzbach. 2001. Self-reported exposures and their association with unexplained illness in a population-based case-control study of Gulf War veterans. *Journal of Occupational and Environmental Medicine* 43(12):1041-1056.

Tango, Toshiro, Toshiharu Fujita, Takeo Tanihata, Masumi Minowa, Yuriko Doi, Noriko Kato, Shoichi Kunikane, Iwao Uchiyama, Masaru Tanaka, and Tetsunojo Uehata. 2004. Risk of adverse reproductive outcomes associated with proximity to municipal solid waste incinerators with high dioxin emission levels in Japan. *Journal of Epidemiology* 14(3):83-93.

Tepper, A., G. W. Comstock, and M. Levine. 1991. A longitudinal study of pulmonary function in fire fighters. *American Journal of Industrial Medicine* 20(3):307-316.

Tornling, G., P. Gustavsson, and C. Hogstedt. 1994. Mortality and cancer incidence in Stockholm fire fighters. *American Journal of Industrial Medicine* 25(2):219-228.

Unger, K. M., R. M. Snow, J. M. Mestas, and W. C. Miller. 1980. Smoke inhalation in firemen. *Thorax* 35(11):838-842.

Unwin, C., N. Blatchley, W. Coker, S. Ferry, M. Hotopf, L. Hull, K. Ismail, I. Palmer, A. David, and S. Wessely. 1999. Health of UK servicemen who served in Persian Gulf War. *Lancet* 353(9148):169-178.

Vena, J. E., and R. C. Fiedler. 1987. Mortality of a municipal-worker cohort: IV. Fire fighters. *American Journal of Industrial Medicine* 11(6):671-84.

Verret, C., M. A. Jutand, C. De Vigan, M. Begassat, L. Bensefa-Colas, P. Brochard, and R. Salamon. 2008. Reproductive health and pregnancy outcomes among French gulf war veterans. *Biomed Central Public Health* 8:141.

Viel, J. F., P. Arveux, J. Baverel, and J. Y. Cahn. 2000. Soft-tissue sarcoma and non-Hodgkin's lymphoma clusters around a municipal solid waste incinerator with high dioxin emission levels. *American Journal of Epidemiology* 152(1):13-19.

Viel, Jean-Francois, Come Daniau, Sarah Goria, Pascal Fabre, Perrine de Crouy-Chanel, Erik-Andre Sauleau, and Pascal Empereur-Bissonnet. 2008a. Risk for non Hodgkin's lymphoma in the vicinity of French municipal solid waste incinerators. *Environmental Health: A Global Access Science Source* 7:51.

Vinceti, M., C. Malagoli, S. Teggi, S. Fabbi, C. Goldoni, G. De Girolamo, P. Ferrari, G. Astolfi, F. Rivieri, and M. Bergomi. 2008. Adverse pregnancy outcomes in a population exposed to the emissions of a municipal waste incinerator. *Science of the Total Environment* 407(1):116-121.

White, R. F., S. P. Proctor, T. Heeren, J. Wolfe, J. Krengel, J. Vasterling, K. Lindem, K. J. Heaton, P. Sutker, and D. M. Ozonoff. 2001. Neuropsychological function in Gulf War veterans: relationships to self-reported toxicant exposures. *American Journal of Industrial Medicine* 40(1):42-54.

Williams, F. L., A. B. Lawson, and O. L. Lloyd. 1992. Low sex ratios of births in areas at risk from air pollution from incinerators, as shown by geographical analysis and 3-dimensional mapping. *International Journal of Epidemiology* 21(2):311-319.

Wolfe, J., S. P. Proctor, D. J. Erickson, and H. Hu. 2002. Risk factors for multisymptom illness in U.S. Army veterans of the Gulf War. *Journal of Occupational and Environmental Medicine* 44(3):271-81.

Young, I., J. Jackson, and S. West. 1980. Chronic respiratory disease and respiratory function in a group of fire fighters. *Medical Journal of Australia* 1(13):654-658.

Zambon, P., P. Ricci, E. Bovo, A. Casula, M. Gattolin, A. R. Fiore, F. Chiosi, and S. Guzzinati. 2007. Sarcoma risk and dioxin emissions from incinerators and industrial plants: a population-based case-control study (Italy). *Environmental Health: A Global Access Science Source* 6:19.